TEXTURE MEASUREMENTS OF FOODS

TEXTURE MEASUREMENTS
OF FOODS

PSYCHOPHYSICAL FUNDAMENTALS;
SENSORY, MECHANICAL, AND CHEMICAL PROCEDURES,
AND THEIR INTERRELATIONSHIPS

Edited by

AMIHUD KRAMER

(Professor of Food Science)
University of Maryland, College Park, Maryland

and

ALINA S. SZCZESNIAK

(Senior Research Specialist)
General Foods Corporation, White Plains, New York

in collaboration with selected authorities on the physiology, anatomy, chemistry, rheology and psychology of texture assessment of food

D. REIDEL PUBLISHING COMPANY

DORDRECHT - HOLLAND / BOSTON-U.S.A.

Library of Congress Catalog Card Number 72–93271

ISBN-13: 978-94-010-2564-5   e-ISBN-13: 978-94-010-2562-1
DOI: 10.1007/978-94-010-2562-1

Published by D. Reidel Publishing Company,
P.O. Box 17, Dordrecht, Holland

Sold and distributed in the U.S.A., Canada, and Mexico
by D. Reidel Publishing Company, Inc.
306 Dartmouth Street Boston,
Mass. 02116, U.S.A.

TABLE OF CONTENTS

PREFACE

Even before the publication of Special Technical Publication 433 of the American Society for Testing and Materials, it became obvious that the brief treatment given to the principles and techniques for sensory measurement and analysis of texture in that volume was all too brief; hence, a task force of ASTM Committee E-18 was formed to develop an authoritative and comprehensive volume on this most complex and important subject to provide within one cover for the student, researcher, and the food manufacturer, a definition and an understanding of the subject of food texture, as well as sensory and objective methods for its measurement. This most difficult task appeared to be possible only after the task force had obtained the assistance of specialists in the many disciplines involved, and after deciding to limit the dissertation to the measurement of texture of foods only.

The task was further clarified when Dr. Finney proposed an outline of six chapters, beginning with one on definition. The second chapter was to be on principles of sensory evaluations, the third on sensory measurements, the fourth on principles of objective evaluation, the fifth on objective measurements, and the final concluding chapter on subjective-objective analogues. The first drafts of these six chapters constituted a symposium on texture presented before a joint session at the 1971 Annual Meeting of the Institute of Food Technology and the American Society for Testing and Materials. Later, it became evident that these manuscripts required substantial elaboration, and three additional subjects were required to complete the publication. These were: structure and texture (Chapter V); indirect (non-mechanical) methods (Chapter VII); and psychophysical measures of texture (Chapter VIII). It was also felt that the concluding chapter could not truly be called 'Sensory-Objective Analogues' so that the treatment and the title of Chapter IX was changed to essentially an essay on statistical procedures for developing objective methods.

Although each chapter bears the name of the author or authors who undertook the responsibility of preparing the first draft, all members of the task force contributed to practically all of the chapters by reviewing and contributing additional material. Special acknowledgement is due to Dr. Malcolm Bourne for undertaking the mean task of integrating the hundreds of references into one list which appears at the end of the volume. Special acknowledgement is due to Mr. W. H. Danker, former Chairman of Sub-Committee II of Committee E-18 of ASTM for initiating this effort and encouraging its accomplishment. Other members of the task force who participated in meetings and assisted in many other ways, but who are not listed as co-authors are: Mr. Fred Dunn, Consumer & Marketing Service, U.S. Department of Agriculture;

Dr. Robert Marvin, Rheology Section, National Bureau of Standards; Professors Mohsenin and C. T. Morrow, Department of Agriculture Engineering, Pennsylvania State University.

We also gratefully acknowledge the permission of Mr. John Klis, Editor of *Food Technology* and Drs. Szczesniak and Sherman, Editors of the *Journal of Texture Studies* for permission to use material previously published by some of the co-authors in these publications.

LIST OF CONTRIBUTORS

Judith A. Abbott, Agricultural Marketing Research Institute, A.R.S., U.S. Department of Agriculture, Beltsville, Maryland.

Caj Åkesson, Swedish Institute for Food Preservation Research, Goteborg 16, Sweden.

Malcolm C. Bourne, Department of Food Science & Technology, Cornell University, Geneva, New York.

Birger K. Drake, Swedish Institute for Food Preservation Research, Goteborg 16, Sweden.

Essex E. Finney, Jr., Agricultural Research Service, A.R.S., U.S. Department of Agriculture, Beltsville, Maryland.

John G. Kapsalis, Food Chemistry Division, U.S. Army Natick Laboratories, Natick, Massachusetts.

Amihud Kramer, Food Science Program, University of Maryland, College Park, Maryland.

Howard R. Moskowitz, Pioneering Research Laboratory, U.S. Army Natick Laboratories, Natick, Massachusetts.

Philip Sherman, Department of Nutrition & Food Science, University of London, Kensington, London W.8, England.

Alina S. Szczesniak, Corporate Research, Technical Center, General Foods Corporation, White Plains, New York.

John N. Yeatman, Division of Food Technology, Food and Drug Administration, Washington, D.C.

CHAPTER I

FOOD TEXTURE – DEFINITION, MEASUREMENT AND RELATION TO OTHER FOOD QUALITY ATTRIBUTES

AMIHUD KRAMER

1. Introduction

Certainly many difficulties, misunderstandings, and actual conflicting conclusions can be avoided when terms are defined precisely so that everyone employing the same term in a specific field is referring to the same things. It would appear therefore that the logical beginning would be with Webster's Dictionary. Unfortunately, however, this very useful compendium lists seven distinct definitions, none of them applying specifically to food. Certainly the English language is flexible enough so that the same word or term may have more than one meaning, but for specific scientific use, the term should be defined with precision. The one dictionary definition that appears to have some application is: "The disposition or manner of the union of the particles of a body or substance." Even this definition, however, would seem to apply more accurately as a definition, or a description of the universe, than as a property of food.

It would appear therefore that the food field requires its own definition of the word texture which is not as yet included in Webster's Dictionary.

The problem of defining texture as a major component of sensory food quality arose when during the 1920's there developed a gradual awareness that sensory quality of foods does not consist of a single well-defined attribute, but is a composite of any number of attributes which are perceived by the human senses individually and are then integrated by the brain into a total, or overall, impression of quality. Among the first to recognize the advantage of such an analytical approach to quality evaluation were those in government agencies, particularly the Department of Agriculture, who were responsible for the development of grades and standards of quality for various raw and processed food products (USDA-CMS).

As late as 1940, however, Lee, writing on the "quality determination of vegetables", was reporting exclusively on what we would now consider textural quality. To this day, we encounter not only lay consumers, but food scientists describing a specific attribute and assigning to it the totality of quality. In a recent conversation with a flavor physiologist, for example, I was impressed by his conviction that sensory quality of foods consists of nothing more than 'flavor', with texture included as contributing to flavor quality in some undefined manner that might be described by some term such as mouth-feel. Similarly, a geneticist indicated his total lack of understanding why a new strain of peach of 'excellent quality' was not being produced

A. Kramer and A. S. Szczesniak (eds.), Texture Measurements of Foods, 1–9. All Rights Reserved.
Copyright © 1973 by D. Reidel Publishing Company, Dordrecht-Holland.

and marketed commercially. Upon further interrogation, it was found that he totally ignored the light color and mushy texture of the peach flesh and was referring to quality strictly on the basis of sweetness and fruity aroma.

2. Classification

Smith (1947) was among the first to list more specific properties of quality, as distinct parameters contributing to overall quality. Of the nine parameters listed (size, viscosity, thickness, texture, consistency, turbidity, color, succulence, and flavor), it is interesting to note that no less than five would probably be included under the general term of texture by many food technologists today.

Kramer (1955) proposed that sensory quality of foods, being a psycho-physical phenomenon, should be systemized or classified in accordance with the senses by which the various attributes of quality are perceived by the consumer. He, therefore, classified sensory quality under the three major senses: appearance as sensed by the eye, flavor as sensed by the papillae on the tongue and the olfactory epithelium of the nose, and kinesthetics (borrowed from Crocker, 1947) or texture as sensed by the nerve endings that subserve muscle. Of 61 commodities for which the U.S.D.A. Standards of Quality were listed in the *Canning Trade Almanac* in 1965, only *absence of defects* and *color* were generally included at weights ranging from 0.15 to 0.6 of the total score – both factors listed by Kramer under the *appearance* category. *Flavor* was listed as a component of quality for only 25 commodities, and *texture* for just 4. The other 57 commodities however, were graded on the basis of such terms as *character, consistency, tenderness* or *maturity*, practically all of which may be classified under the general (or primary) term of texture.

At that time, therefore, there was some reluctance to assign the entire area of kinesthesis (and/or haptaesthesis, Muller, 1969) to texture, since other terms – such as viscosity and consistency – were also in general usage, particularly in official grades and standards. The fourth sense involved in quality evaluation of food, namely sound as sensed by the ear, was mentioned occasionally – but usually incidentally – as being only of minor importance in the overall evaluation of food quality.

Another reason for reluctance on the part of some workers to assign a primary role to the term texture in sensory evaluation of food, was its ubiquitous use so that its precise meaning was not immediately evident. Judging by the usual standard dictionary definitions, the term texture was first used in the textile industry in connection with the art of weaving, as "disposition or connection of threads as in a fabric" (Webster's Dictionary). It is only in the last decade or so, with the development of 'textured foods', that this original meaning of texture could be applied directly to the evaluation of food quality. A more broadly applicable definition such as "the disposition or manner of union of particles of a body or substance", could also apply to all natural and processed 'solid' foods, where "the disposition or manner of union" of different types of cells and tissues in the food material could be considered as the 'texture' of the food. This, therefore, is a major attribute of food quality which can be included under a definition of texture and which would be directly related to the internal structure of natural or fabricated foods (Sherman, 1972).

Webster's Dictionary also lists a definition of texture specifically for petrography as "smaller features of a rock – granular" which has been and still is applied in some food areas, as for example, in meats, where the smoothness or coarseness of the muscle fibers is of concern. Thus here again structure is involved, but in this instance it would be difficult to call this characteristic as something pertaining to the sense of feel only since it is obviously noticeable to the sense of sight. Certainly the property of 'marbling', that is the interlacing of fat within the muscle, could well be included under one of the dictionary definitions of texture, but here again it could as well or more readily be classified under the general term of appearance.

Viscosity and consistency, on the other hand, were classified by Kramer and Twigg (1959) as appearance factors, since they were generally used in grades and standards of quality as referring to the flow of liquid and semisolid drinks or slurries which could be readily perceived by the eye of the observer before his kinesthetic sense came into play. Thus, in their book on *Quality Control for the Food Industry*, Kramer and Twigg (1970) treated 'viscosity and consistency' under the general heading of appearance, and texture alone under kinesthetics. Viscosity and consistency were treated in accordance with classic rheological concepts as applying to Newtonian and non-Newtonian liquid and semi-solid materials respectively*, while the term texture was reserved for 'solid' foods whose complex structure caused difficulties in converting sensory responses in terms of classic rheological model systems.

Kramer therefore proposed to confine texture further – from the sensory standpoint to the sense of feel only, and from the physical standpoint to the part of rheology that deals with the deformation or flow of matter, but only as a result of the application of

	Rheological or physical terms (gravitational force)		
Psychological or sensory terms	Up to 1.0 gravity		Greater than 1.0 gravity
	Newtonian	Non-Newtonian	
Sight - - flow or spread	VISCOSITY	CONSISTENCY	
Feel - - mouth or finger			TEXTURE
Taste and smell	- FLAVOR -		

Fig. 1. Classification of texture in relation to force required to initiate flow (Kramer, 1964).

* It should be noted that the British Dairy Industry usage of *consistency* is synonymous to this restricted definition of *texture* (Scott Blair, 1968). Also note the similarity of treatment for fluid systems by Corey and Creswick (1970).

forces greater than gravity (Figure 1). Thus 'gel strength' was considered a textural characteristic since a force greater than 1.0 g is required to cause deformation and it is sensed primarily by mouthfeel. On the other hand, consistency of a sauce was not included, since its flow characteristics may be observed by the sense of sight, and flow occurs with the application of gravitational forces alone. He further suggested that which food slurries such as sauces, being invariably non-Newtonian, this characteristic could properly be called 'consistency'. Flow characteristics of an oil or syrup were termed 'viscosity' since they also deformed under the force of gravity alone and, in addition, being quite homogeneous and in many instances true solutions, were more likely to exhibit Newtonian flow.

Separate parameters of texture were further classified in accordance with the nature of the force applied to cause specific types of deformation such as compression, shearing, extrusion, or a combination of these.

During the decade of 1960, a number of additional textural classifications were proposed (Szczesniak, 1963b; Bourne, 1966) which, while dealing largely with 'solid' products, did not specifically exclude liquid or semi-liquid foods, while Sherman (1969) utilized the state (solid, semi-solid, fluid) of the product in his classification of the mechanical (masticatory) properties. Considerable progress was also made, notably by Mohsenin (1970), in defining mechanical properties of solid foods in rheological terms, thereby providing a much more precise understanding of the specific mechanical properties that could be related to very specific human kinesthetic responses.

3. Definition

With the advent of the *Journal of Texture Studies* in 1969, the term texture may be considered to have been generally accepted as a major division of sensory quality covering all kinesthetic responses of foods in whatever state they are in. By the same token, in distinguishing texture from the other major sensory categories, it may be appropriate to relegate characteristics such as visual appearance or particle size (Sherman, 1969) to appearance as well. Other attributes such as sensations of cold, heat, or oiliness would be classified under flavor as well as texture. Our definition of texture may further be limited to sensations of touch or feel by the human hand and mouth parts, since quality attributes of food are not ordinarily sensed by other human organs prior to or during mastication.

Any definition, and one of texture is no exception, must be arbitrary especially in drawing borderlines. It should be recognized that every primary quality attribute (as well as secondary attributes) is not entirely independent and may overlap and certainly be influenced by other attributes. Thus Kramer (1968) portrayed sensory quality in the form of a finite continuum (Figure 2), or circle, with the primary attributes of appearance, texture (kinesthetics), and flavor as sharing the periphery of the ring. At one part of this periphery between appearance and texture there is an overlapping zone where terms such as consistency and viscosity may be placed, since these can be classified under both appearance and texture. At the other side of

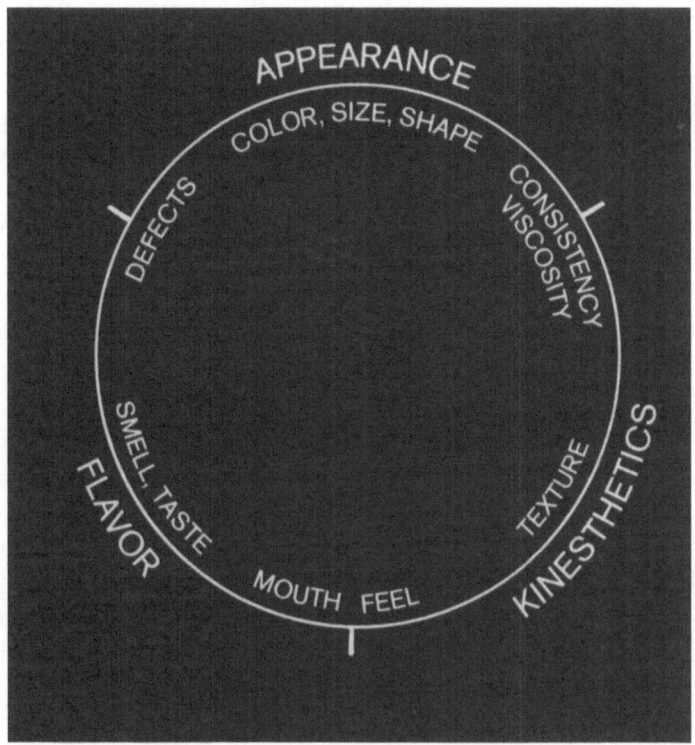

Fig. 2. A schematic presentation of sensory quality of foods as a finite continuum (Kramer, 1968).

the zone where texture meets flavor, there is a similar overlapping where the term mouthfeel may be placed.

Accepting the inevitability of such overlapping borders, there is nevertheless general agreement that the term texture rather than kinesthesis or haptaesthesis *is* the accepted popular term of *that one of the three primary sensory properties of foods which relates entirely* (or in addition to the other primary properties) *to the sense of touch or feel* and is, therefore, at least potentially capable of precise measurement objectively by mechanical means in units of mass, or force. The equivalent psychological and physical terms are kinesthesis (the muscle sense), or haptaesthesis (the skin sense), and rheology (the deformation or flow of matter), respectively.

4. Measurement

4.1. SENSORY QUALITY AND SENSORY MEASUREMENT

A sensory property is one that is perceived by one of the senses with which the food consumer evaluates the product. Hence, texture is a sensory property of foods. Its measurement may be accomplished directly by the use of the respective sense (in this case the sense of touch) and in such instances the test is subjective (Stevens, 1966b).

Although texture is a sensory property, there is the opportunity of measuring it by physical, more specifically rheological objective methods. In general, objective techniques for measuring sensory properties have the basic disadvantage of measuring sensory properties only indirectly and are, therefore, accurate only to the extent that they are analogous to the human sensory response. At the same time they have the advantage of objectivity in that – at least potentially – they are not as subject to drift, fatigue and are more precisely calibratable than human sensors.

Even less direct, non-rheological, objective methods have been used successfully for measuring textural properties of foods, particularly where the sensory quality is not specifically defined, but is a broad concept based largely on the physiological condition of the plant or animal tissues constituting the food (e.g. maturity). Such non-rheological measurements may be colorimetric, densimetric, or determinations of chemical composition. Success of such measurements depends not so much on the equivalency, but on the coincidental relationship that the test results happen to change consistently with changes in consumer responses for the quality attribute. Thus although such procedures may be successful in many instances, they may suddenly become unsatisfactory. Maturity of lima beans for example was measured adequately for many years by the lightness of the product, since lima beans, as they mature, lose chlorophyll which provides them with green pigmentation. This method was successful until a new variety was developed which retained its green coloration as it matured. Consequently large quantities of hard, mature undesirable lima beans were graded as young, immature succulent lima beans, simply because they retained greenness and did not turn 'white' on maturity (Kramer and Hart, 1954).

The lima bean problem was solved simply by replacing the color (lightness) test with a physical test of hardness, similar to the 'tenderometer' test for raw peas. This new hardness test was satisfactory, since the specific parameter to be measured was hardness, and fresh lima beans and peas became harder as they increased in maturity. For canned peas however, the hardness test was not satisfactory. In fact, tender, less mature raw peas remained firmer upon canning than more mature peas. Furthermore, any lot of pea of any maturity could be 'softened' to any desired level simply by extending the time and temperature of heat processing. For the canned peas, however, the sensory property to be tested was not specifically that of hardness, but a combination of crispness and lack of mushiness, which was more accurately determined by a chemical measurement of the alcohol insoluble solids, largely starch, plus a mechanical test for hardness (Angel *et al.* 1965).

Szczesniak has recently prepared a comprehensive review of instrumental (1973a) and indirect (1973b) methods of texture measurement. She lists the basic elements of such instruments as being: (1) a probe contacting the food sample; (2) a driving mechanism for imparting motion (and stress); (3) a sensing element for detecting the resistance of the foodstuff (strain); and (4) a readout system. She further classified the types of texture measuring devices as: (1) penetrometers; (2) compressimeters; (3) shearing devices; (4) cutting devices; (5) masticometers; (6) consistometers; (7) viscometers; (8) extrusion measurements; and (9) multi-purpose units. The indirect methods are classified as: (1) chemical; (2) enzymatic; (3) microscopic; and (4) physical.

4.2. ACCURACY OF OBJECTIVE METHODS

At the risk of redundancy, it may still be appropriate to point out at this time that objective measurements are unquestionably useful and in many instances preferable to subjective measurements in the evaluation of rheological-mechanical properties of new fabricated food products, but their use in the evaluation of complex natural products may be limited. Liquid foods exhibiting Newtonian characteristics such as some oils, syrups, and juices, may be measured objectively and accurately and reported in generally acceptable rheological terms. Non-Newtonian liquids or semi-solids, and particularly natural 'solid' foods, may be more difficult to measure objectively with sufficient accuracy, and for this reason any such proposed objective method should first undergo thorough examination and validation to establish its accuracy before it is approved for routine use. A generalized procedure for accomplishing this was suggested by Kramer (1956).

(1) Definition of parameters to be measured – the more precisely the individual parameters are defined, both in sensory and in physical terms, the greater the probability of developing a successful objective test. This is essentially a search for psycho-physical analogues (or sensory-rheological analogues).

(2) Statistical analysis for selecting the best objective method or combination of methods for each sensory parameter by use of correlation-regression analyses, where sensory evaluation of texture is always the dependent variable (y) and the objective test value is always the independent variable (x); thus the regression equation assumes the form:

$$y = a + \beta x.$$

Since relationships between subjective and objective values are usually non-linear, a truer and higher correlation can frequently be obtained by converting sensory data to their logarithms (Hopkins, 1950; Stevens, 1966a). Where relationships are more complex and such a simple expedient will not suffice, there may be a need for adding higher order functions (e.g quadratic, cubic, etc.). Thus the regression equation will appear as:

$$y = a + \beta x + \beta^2 x^2 \cdots + \beta^n x^n.$$

Where more than one independent variable (objective test) is involved, the simplest form of the regression equation will be:

$$y = a + \beta_1 x_1 + \beta_2 x_2 \cdots \beta_n x_n.$$

Higher order functions may be added as required.

By such analyses, those measurements that contribute significantly to the predictability of the sensory parameter will be retained while others will be dropped. The partial regression coefficients (β) will also indicate the relative importance of each of the objective methods while the coefficient of determination (R^2) will indicate the extent to which the selected method or methods are capable of predicting the sensory quality level of the parameters studied.

(3) Validation – since there is always the chance that the set of samples on which the original studies were made yielded results that may be limited to little more than chance relationships among these samples only, it is necessary to validate the accuracy (or predictability) of the selected procedure. This may be accomplished by repeating the study with additional samples that cover but do not exceed the usual commercial range for each quality.

(4) Integration of selected objective scales for predicting overall textural quality. In the above mentioned search for objective methods for measuring specific parameters, some methods may have been selected that may contribute significantly to the prediction of more than one specific parameter of texture. Thus, by means of multiple regression and other multivariate analyses (Kapsalis *et al.*, 1973) it may be possible to reduce the number of objective tests that will provide adequate predictability for overall textural quality.

It is only after such elaborate studies that an indirect, objective method or a combination of methods can be developed to predict with sufficient accuracy a sensory quality such as texture. A common flaw in the development of these procedures is that the 'validation' step (3) is either omitted or not performed thoroughly, so that in application, under conditions which are always subject to change, gross errors may be made in assigning quality levels to specific lots of given commodities simply because variations in certain variables were not encountered during the studies leading to the procedures used. The reverse may also be true – results obtained from objective tests performed with certain instruments may be taken as the true sensory quality levels when in fact the consumer is indifferent or unaware of the differences indicated.

Another word of caution may be called for, and that is to use an adequate number of samples. With the rapid proliferation in available equipment on one hand, and improved training and precision of taste panelists on the other, large numbers of dependent and independent variables may be defined, measured and analysed statistically. With the availability of high speed computers, the statistical analysis is usually not a serious problem. There is the risk, however, that one may ignore the fact that with the addition of every variable (i.e. parameter, and/or test procedure) we lose one degree of freedom. Thus, if the number of test procedures entered into a multiple regression analysis approaches or exceeds the number of samples, we are left with no degrees of freedom. This inevitably leads to a conclusion that the combination of objective tests predicts perfectly the sensory parameter. This, however, is nothing more than a meaningless statistical exercise, since a negative number of degrees of freedom must result in a perfect fit (Kramer, 1966; Rasekh, 1968).

4.3. PRECISION OF OBJECTIVE MEASUREMENTS

While precision (i.e. reproducibility) of objective measurements is potentially superior to sensory, this cannot be taken for granted. Thus, an imprecisely manufactured instrument can result in as great or greater inter or intra instrumental error than an error arising as a result of differences in human responses particularly those obtained from trained panelists. It is, therefore, important that instrumentation for objective measurement of sensory properties be produced with the utmost precision, and the tests performed under carefully controlled and specified conditions. Preparation of the food sample and its loading in the instrument must also be rigidly specified and precisely performed.

Size and shape of the test cell as well as the test sample must be rigidly specified and maintained. Thus for example, enclosing the sample in a container so that deformation can occur in only one direction could simplify substantially the nature of the forces employed in the test, thereby substantially reducing the testing error.

In general, the larger the sample units or number of units in the sample to be tested, the more precise and efficient can the expected results be. Thus for example, the average of 100 individual puncture tests on 100 kernels of corn may be less precise and more time-consuming than one compression-shear-extrusion test on one sample consisting of 400 kernels of corn.

Given a thorough understanding of the specific problem involved, following the performance of a thorough and adequate study, there is no reason why the texture of any product cannot be measured accurately and precisely with the appropriate instrument or instruments. In fact, it is frequently found that as few as 3 or 4 instrumental measurements can predict adequately all the textural properties of a food product; in some instances where a specific parameter is dominant (as hardness of peas) only one test may suffice. This is an indication that the objective measurement of textural quality may well be far less complex than the objective measurement of other sensory properties such as odor.

CHAPTER II

PHYSIOLOGICAL ASPECTS OF TEXTURE PERCEPTION, INCLUDING MASTICATION

JOHN N. YEATMAN and BIRGER K. DRAKE

1. Introduction

In a review of physiological aspects of food texture perception, we should remember that the senses of sight, touch, and hearing, together with other senses interact to give a pattern of multiple responses.

Once we accept a food visually and begin to consume it, we bring into play the senses of taste and smell. Sensory perception of texture, however, depends mostly on deformation resulting from the application of pressure and/or on surface properties such as roughness, smoothness, or stickiness estimated by the sense of touch. While a consumer may develop an idea of the texture of a food when he handles it, texture is indicated best by the sensations caused by contact with hard and soft parts of the mouth.

Mastication has 3 physiological functions: to break up and lubricate food so that it can be easily swallowed; to mix it with salivary enzymes; and to increase the surface area so that the food may be more quickly attacked by gastric secretions.

One objective of this chapter is to elucidate how the physiology-psychology of mastication depends on local actions induced by mechanical, chemical, and thermal agents, and how everything interacts to enable us to decide on the textural quality of a product. It is, in this context, of interest to consider the idea of Pierson and LeMagnen (1970) who claimed that attempts to measure physical characteristics of foods and to correlate these measurements with sensory evaluation have not paid enough consideration to the physiological processes of mastication (chewing) and deglutition (swallowing) involved in the generation, by food texture, of an oral spatio-temporal pattern of sensory stimulation.

2. Physiological Basis of Mastication

2.1. GENERAL

Once biting and chewing start, a very complex feed-back pattern of stimulation and motoric action is set up. Kawamura (1964), in his extensive paper on the physiology of mastication, paid particular attention to neurophysiological characteristics. He suggested that a special separate neural system provides a coordinated function between the muscles of jaw and tongue. The mouth-jaw joints on the right and left sides of the oral cavity function simultaneously; therefore, the functional relationships of the masticating muscles are somewhat different from muscles controlling the extremities. It can be added that the process of mastication needs not only the teeth but also the lips, cheeks, tongue, palate, salivary glands, and all other oral structures to prepare food for swallowing.

The tongue, playing a very important part in mastication, moves the food into the correct position for chewing. In fact, it may exert a direct crushing effect by pressing soft food against the hard palate. Mixed with saliva the food is finally pushed toward the back of the mouth by the action of the tongue. The tongue responds to its sensory modalities such as taste, touch, and temperature, which play an important role in the feed-back mechanisms that influence movements of the tongue and jaw.

Another valuable review on mastication and other oral physiological factors has been published by Kapur *et al.* (1966). Both the considerations by Kawamura (1964) and those of Kapur *et al.* (1966) are a strong confirmation of the above mentioned idea of Pierson and LeMagnen (1970).

2.2. FORCES AND DEFORMATIONS OCCURRING IN THE MOUTH

There is no doubt that the teeth play a considerable part in the appreciation of texture and geometrical properties of substances in the mouth. Anderson (1953) made oscilloscope records of stresses acting on teeth during mastication (Figure 1). Stress measurements were accomplished by inserting a strain gauge into a special inlay (Figure 2). The tracings (reading the curves from right to left) for biscuit show greater force on teeth (in excess of 8 kg) which lasts for a shorter time before the bolus is ready for swallowing. The tracings for carrot and meat, as contrasted to biscuit, show in the first few seconds of mastication a relatively great force (6 kg or lower) which more gradually diminishes to a low stress value, but which requires a considerably longer period of time to prepare a bolus suitable for swallowing. By applying a known static force for 2 s laterally to a tooth, Mühlemann (1960) was able to show, in some 10000 measurements, that the movement of both single- and multi-rooted teeth oc-

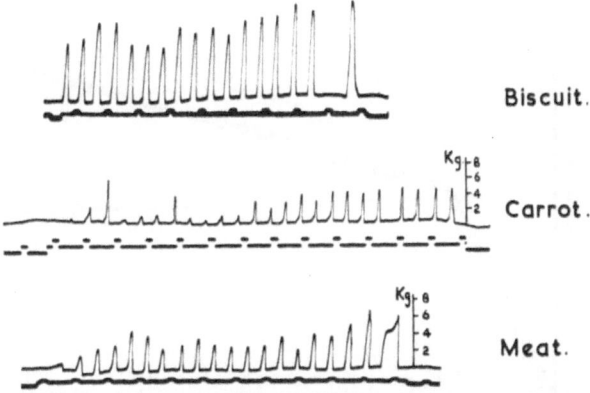

Fig. 1.   Oscilloscope records of stresses on teeth during mastication of three foods. Lower tracing in each record shows elevations at one second intervals. (After Anderson, 1953.)

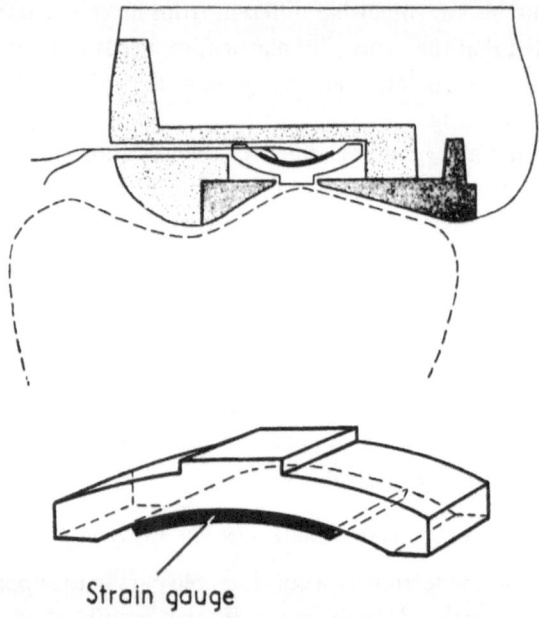

Fig. 2. Method for measuring occlusal pressure during normal eating. The strain gauge (shown enlarged below) is inserted into an inlay in the tooth. (After Anderson, 1953.)

curred in 2 phases (Figure 3): rapidly increasing when the force was increased to 100 g and slowly increasing when it exceeded this figure. A point was reached at a force of 1500 g at which no further movement occurred and pain was felt.

Data reported by Worner and Anderson (1944) suggest that the effect of practice, such as that performed by the habitual gum chewer, can increase the biting force (Figure 4). Knowledge of the habits of individual subjects used in panels could enable a

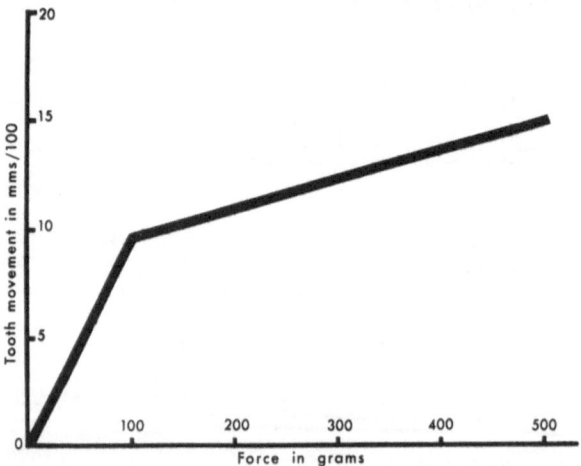

Fig. 3. Effect of the application of increasing force on tooth movement. (After Mühlemann, 1960.)

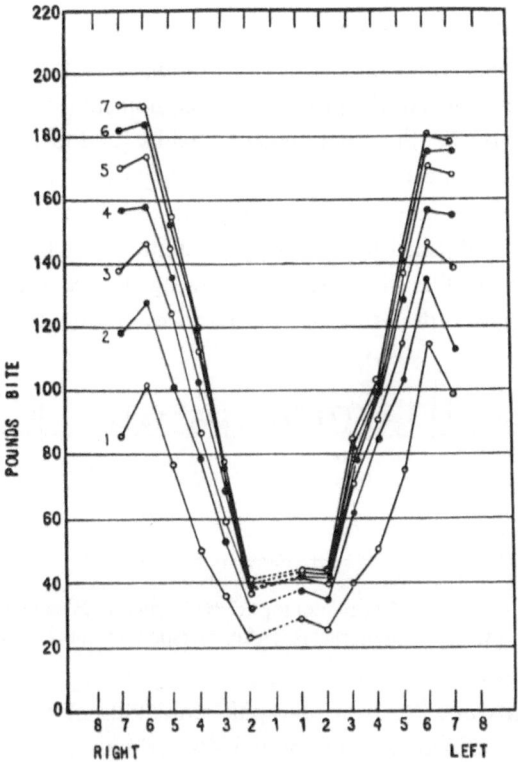

Fig. 4. Records of biting forces of individual teeth. Curves 1 to 7 show effect on the strength of bite of 3, 4, 7, 9, 11, and 14 days' practice respectively. (After Worner and Anderson, 1944.)

more powerful analysis of data whereby, e.g., the correlation between objective and sensory measurements might improve.

An engineering approach to the evaluation of forces of mastication has been taken by Ledley (1971) who in a recent article presented a mathematical treatment of such forces.

Anderson and Picton (1957), experimenting with 10 people, recorded simultaneously contacts between the teeth and electro-myograms of the masseter muscle. They found that the proportion of strokes in which the teeth came into contact varied for different foods, e.g., with bread or meat about one-half of the subjects made contact with every chew, but with biscuits the proportion was much less. Jankelson *et al.* (1953) found that food was between the teeth most of the time throughout mastication and that the teeth contacted very infrequently but did contact during swallowing. A more recent investigation of jaw movements was performed by Bewersdorff (1969). Many more studies of the kinds referred to above have been reported in the odontological literature but cannot be enumerated here.

Vibrations which occur when more or less crisp foods are chewed have been studied by Drake (1965a, b) and Kapur (1971), who used specially adapted sound recording and analyzing techniques. Such studies can provide information which is not obtainable by simple force-deformation measurements.

2.3. CHEWING-SWALLOWING PATTERNS

To study the complex space-time pattern of oral stimulations, Pierson and LeMagnen (1970) recorded the responses of the physiological equipment (the human mastication apparatus) directly involved in the pattern generated by food textures rather than the responses of an imitating mechanical equipment. They described electromyographic recordings of the masseter muscle using skin electrodes fixed on each cheek at the suitable motor point established by electrical stimulation. At the same time deglutition was detected by recording the noise of swallowing. Typical recordings are shown in Figure 5. The edogram (derived from the Latin *edere*, to eat) shows tracings of

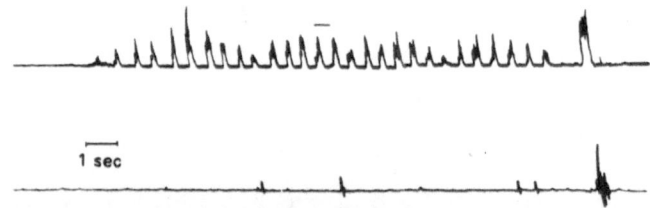

Fig. 5. Edogram of a bolus of food (chocolate) top: electromyographic potentials of masseter muscles, bottom: swallowing movements. (After Pierson and LeMagnen, 1970.)

swallowing movements and electromyographic potentials of the masseter muscles when a bolus of food (defined as a soft, round-shaped mass of foodstuff undergoing mastication) is consumed. Many rapid chewing movements are required to consume the chocolate while only a few swallowing movements are necessary to prepare the food for deglutition.

Obviously, many methods mentioned in Chapter III are also capable of giving information on chewing patterns.

2.4. TACTILE SENSE ORGANS AND THEIR FUNCTION

Pfaffmann (1939) recorded impulses in the dental nerves of the cat and showed that the principal sensory receptors were almost certainly in the periodontal membrane which surrounds the tooth in its socket in the jaw. Later work has shown that, when the tooth is slightly displaced by pressure (cf., above, Mühlemann, 1960), the resulting deformation, probably a compression of the periodontal membrane, is the stimulus detected.

As Oldfield (1960) suggested, the sense organs involved in the mechanism of mastication may be grouped: (1) those in the superficial structure of the mouth (hard and soft palates, tongue and gums); (2) those around the roots of the teeth; and, (3) those in the muscles and tendons. The hard palate has a practically unbroken cellular surface highly sensitive to touch; coarseness of food is detected mostly by receptors in this surface. The muscles and tendons may play a relatively small part in the appreciation of texture and are chiefly concerned with regulating the muscular contractions in a coordinated and properly phased fashion.

Gairns (1951), Gairns (1954), and Jenkins (1966) have shown that the soft structures of the mouth are provided not only with a network of free nerve-endings, but also with a variety of organized terminations, encapsulated and unencapsulated, varying in type but on the whole resembling those found elesewhere in other mucous membranes and in smooth skin; however, nobody has determined the number of basically different forms, nor in what way differences in structure are correlated with differences in the kind of sensation they subserve.

3. Psychological Implications

3.1. RELATIONS BETWEEN INSTRUMENTAL AND SENSORY DATA

Aspects of sensory evaluation of foodstuffs, and also correlations between instrumental and sensory properties, are treated elsewhere in this volume. Here, only the following will be mentioned.

From experiences reported in other fields of psychological inquiry, such as color science, it may be wrong to expect too much from impressions and preferences of individuals about food texture, however well-standardized the experimental procedures, and well-designed the statistical analysis. This does not mean that we should refrain from sensory analysis. On the contrary, panels are often indispensable, but their use should be made more effective by learning more about the physiological and psychological mechanisms involved.

Sophisticated instrumentation and/or computerization have relieved us of many tedious tasks (and their worth is not questioned here), but they do not necessarily lead to a grasp of the qualities implied. Harper (1968) speculated whether phenomena falling within the scope of pattern perception remain outside the present possibilities of Computer Methods; in any case, however, the most serious problem at present is the lack of adequate mathematical models for coping with the complexities of perception processes. Another way of expressing this is to say that, even aside from purely social aspects, the human can learn to perceive qualities which cannot be expressed in analytical terms, at least not without lengthy and expensive research. Also, while instrumental measurements are aimed at replacing traditional forms of sensory evaluation, it must be remembered that mathematical-statistical procedures, like multiple regression, are descriptive rather than interpretative; therefore, predictions from instrumental data to sensory properties demand that such procedures be supplemented with a thorough knowledge of underlying mechanisms. All this is important also in the field of texture studies.

3.2. EATING HABITS

As Oldfield (1960) concluded in his above mentioned paper:

What, on any given occasion, an individual or set of individuals may report as to his likes and dislikes, or as to his impressions of sensory quality, may in itself reveal little of the factors at play or of his future tendencies to take this or that foodstuff. To gain some insight into these, and provide thereby a firm basis for practical application, it is essential first to sort out the constituents of the sensory and motor functions involved.

The activity which brings about perception of the texture and other properties of substances in the mouth is to some extent an automatic one, and all of the sensory activity does not necessarily penetrate into consciousness. This means that, apart from the associated emotional reactions, responses are often evoked at a fairly involuntary, physiological level; in every case, however, psychology may be involved in a way which is as yet only dimly recognized. One behavior may consist of rejection of the food *or* automatic initiation of the next stages of ingestion. Another behavior may reflect a state of hunger or a need for protection against substances which are detrimental to the individual. A basic biological mechanism is related to age, past experience, and present organic state of the individual, while another mechanism, influenced by psychological factors, involves short or long tendencies such as fashions or crazes.

Some foodstuffs may induce aversion to further chewing and swallowing, which suggests that perhaps mastication is dependent for its continuance on stimuli acting on both touch and taste receptors.

CHAPTER III

SENSORY ASSESSMENT OF TEXTURAL ATTRIBUTES OF FOODS

JUDITH A. ABBOTT

1. Introduction

People of higher educational, social, or economic status and those exposed to a greater range of foods are generally more aware of food texture than less cosmopolitan people (Szczesniak and Kleyn, 1963). As the consuming public becomes more conscious or critical of food textures, for instance with increasing use of processed and synthetic foods, the need for better methods of measuring and controlling texture grows. Instruments can be devised to measure many food attributes, but these measurements are of little value unless they relate to consumer evaluations. The food industry, therefore, needs reliable sensory panels for three reasons: (1) to assess the relative importance of texture to the acceptability of a food item – *whether to measure texture*, (2) to determine the textural characteristics which are important in that food – *what to measure*, and (3) to evaluate the appropriateness of a particular objective test for a textural characteristic – *how to measure*. According to Kramer (1968):

An objective method cannot possibly be an improvement in accuracy over a subjective method, since the accuracy of an objective procedure can be determined only by its degree of correlation with the subjective evaluation sought. The precision and calibratability of the objective method, however, can be superior since a physical or chemical procedure should provide closer duplication than a human sensory panel, and by use of references or blanks, objective methods should be more calibratable than human judges who are always subject to drift.

However, correct panel methods can greatly improve the reliability and validity of sensory tests.

Food scientists frequently decide whether texture is important in a given food on the basis of intuition, interest, or experience. There is a need for further examination of consumer awareness of and sensitivities to textural attributes in various foods. Recently, papers on the use of texture-describing words for specific foods based on consumer interviews have been published. Harper (1968) mentions that a compilation of terms called the 'Language of Flavor' begun in 1956 contains about 60 words better classified as textural terms. Many of the words in that list and in the terminology surveys mentioned in the following section of this paper are ill-defined, and may have far different connotations for different individuals. Such a personal experience as mouthfeel is difficult to define precisely, but it is necessary to have agreement on the meanings of the terms to have useful sensory panel results.

2. Terminology Surveys

Szczesniak and Kleyn (1963) and Szczesniak (1971) investigated the relative impor-

A. Kramer and A. S. Szczesniak (eds.), Texture Measurements of Foods, 17–32. *All Rights Reserved.*
Copyright © 1973 by D. Reidel Publishing Company, Dordrecht-Holland.

tance of texture to consumers and some of the texture terms which they use. Unrestricted word-association tests were conducted by interviewers who asked the respondent to supply the first 3 word associations which came to mind upon hearing a particular food mentioned. The list of foods included beverages, meats, baked goods, fruits, vegetables, desserts, and snacks. In the first test, technical and nontechnical men and women at General Foods Technical Center were interviewed. Since people working for a food company may have unusual awareness of food attributes, a further test was conducted with typical consumers. Results indicated little difference between the two groups. The greatest number of responses were classed as menu uses, including methods of preparation, components, accompanying foods, and serving occasions. The category receiving the next most responses was food attributes, particularly texture, flavor, and color. Technical personnel gave nearly as many food attribute responses as menu uses. Texture received slightly more emphasis than flavor from the General Foods group, slightly less from consumers. The texture terms most frequently given were crisp, dry, juicy, soft, creamy, crunchy, chewy, texture, smooth, stringy, hard, and flaky.

In a study by Yoshikawa *et al.* (1970) in Japan, college women were asked to select the pertinent words from a list of 40 texture terms and indicate their relative importance for each of 97 food items. The majority of the foods were Japanese and many of the words are not translatable so their results are not applicable in the United States. However, the weighted word-association test appears to be valuable in establishing the importance of various textural characteristics of foods to the intended consumers. Careful selection of the texture terms is essential.

It is necessary to establish the important textural attributes for each type of food examined and to have complete agreement on the *definitions* of the attributes. Stier (1970) reported that a relatively expert panel concluded that potato chip quality could be satisfactorily rated using the following 3 terms as defined: *crisp* – initial sensation of being brittle, crushable, or friable; *crunchy* – harsh sound produced when chip was broken and chewed; *abrasive* or *rough* – sharp granular pieces in mouth after first chew. With such definitions, any person not in the original panel could quickly understand the terms and, after sampling potato chips with a range of qualities, could evaluate them using the same three terms. The same terms could be given other sensory definitions or might require only a shift of intensity scales to reflect the quality of other food types, such as crackers or raw fruit or vegetables. The *Better Homes and Gardens Encyclopedia of Cooking* (1970) uses the terms firm, semi-firm, crisp, and tender to describe the texture of various apple varieties.

The texture profile discussed in Section 7.4 of this chapter also utilizes many popular texture terms after deriving specific definitions for them.

3. Consumer Methods of Nonoral Texture Evaluation

To learn the nonoral sensory methods used by consumers to assess food firmness, Szczesniak and Bourne (1969) used a *paired comparison* test. Paired samples of various foods were presented in random order. The two samples of each food were selected to have sufficient textural differences for definite distinction of the firmer sample. The panelists were asked to indicate which of each pair was more firm, using any method they wished except placing the food in the mouth. They were permitted to compare the two samples as much as necessary. Observers noted that panelists utilized four principles in sequence to evaluate texture as firmness increased from very soft to very firm:

> Viscosity/consistency (puddings, whipped toppings)
> Deformation (bread, marshmallows, tomatoes, lettuce, pears, apples)
> Puncture (pears, apples)
> Flexure (carrots)

Stevens and Guirao (1964) tested the ability of people to evaluate fluid viscosity by visual or tactile means only or by combined visual and tactile means. A series of fluids with viscosities from 10.3 to 95000 centipoises was prepared. These were presented to 10 judges for each of the tests. The order was randomized except that the least and most viscous samples were never presented first. For visual judgement, the closed container was turned or shaken and the flow was observed. For the tactile judgement, the judge stirred the fluid with a rod while blindfolded. For the combined visual and tactile evaluation, the judge watched while he stirred the fluid with a rod. The judge assigned the first sample a number and rated the subsequent samples in relation to the first. The geometric means were plotted on a log-log scale. The slopes were approximately 0.5 for all three methods, i.e., the perceived viscosity increased as the 0.5 power of the apparent physical viscosity.

4. Sensory Panel Tests

Sensory panels are groups of people used to evaluate the quality or characteristics of foods or other materials. The kind of people selected for a panel is determined by the question to be answered by the panel. The panel may consist of from 50 to several hundred untrained, randomly selected consumers. Or the panel may include as few as 5 very carefully selected, highly trained people of exceptional sensitivity. Several good sources of information on sensory panels are the ASTM *Manual on Sensory Testing Methods* (1968a), Amerine *et al.* (1965), Larmond (1970), Dawson and Harris (1951), and Ellis (1966).

There are 3 fundamental types of sensory panels with 3 very different functions. These are not unique to texture evaluation and have been described extensively for flavor, odor, and other sensory testing. *Preference/acceptance* panels evaluate the opinions or likes and dislikes of the consuming public and should be conducted using consumers. Preference tests determine which of the two or three samples most people like best. Acceptance tests indicate if a product is likely to be accepted by most potential consumers. *Discriminatory* tests determine whether there are detectable differences among samples. They do not indicate how large or what kinds of differences exist, only if the samples are not identical. *Descriptive* tests indicate the kinds or magnitudes of differences among samples. Most of the sensory panels reported in the literature are intended to be of a descriptive nature because the researcher is interested in the effects of the studied variables on the food.

5. Preference/Acceptance Tests

Which sample do you prefer?
Would you accept this sample?

Preference/acceptance tests measure the opinions of consumers.

Results of preference/acceptance tests are seldom published because the chief aim of consumer testing is to determine salability of proprietary goods. Consumer tests of specific products normally do not concern only the textural attributes but include all attributes which differentiate among the samples or characterize a new product.

Preference/acceptance tests require panelists which accurately represent the population for which a product is intended. In order to represent a large group of consumers, a large panel size is necessary. It is often not economically feasible to use proper consumer tests so a laboratory 'consumer' panel may be used for preliminary screening of samples during product development.

In laboratory situations, it may be necessary to compromise and select panelists at random from the available personnel, excluding any with special knowledge of the product. Panels as small as 16 members are sometimes used, but usually at least 30 are used on laboratory 'consumer' panels. Even a 30-member panel is so small that important trends may not be detected because of the large error factor or the narrow range of sampling. In general, 50–100 panelists are recommended for such laboratory tests (ASTM, 1968).

The experimenter must be careful to interpret the results within the limitations of the test. Laboratory panels should periodically be compared with typical consumer panels to ascertain their validity.

Panelists for preference or acceptance tests should be instructed only in the mechanics of the test. There must be no attempt to influence their attitudes or their manner of arriving at decisions. Such influences are antithetical to the purpose of the consumer tests (ASTM, 1968a).

Probably the most used form of consumer tests is the paired comparison test. A pair of samples is given to an untrained person with a request to indicate *which* sample he likes better. In another test type, the person indicates on a hedonic rating scale such as those in Figure 1 how much he likes the samples. Verbal scales are more customary but the facial hedonic has been found effective where use of verbal scales is limited by language barriers or by the age of participants, such as with children.

Schutz (1965) developed a 9-point 'Food Action Rating Scale' (FACT for *Food Action*) for consumers to indicate whether they would or would not use a product. The criteria may be from 'I would eat (or buy) this every opportunity I had' to 'I would eat (or buy) this if I were forced to'. According to Schutz the FACT scale means were highly correlated with hedonic scale means, but were always lower and the distribution less skewed. He found the FACT scale more sensitive to food differences.

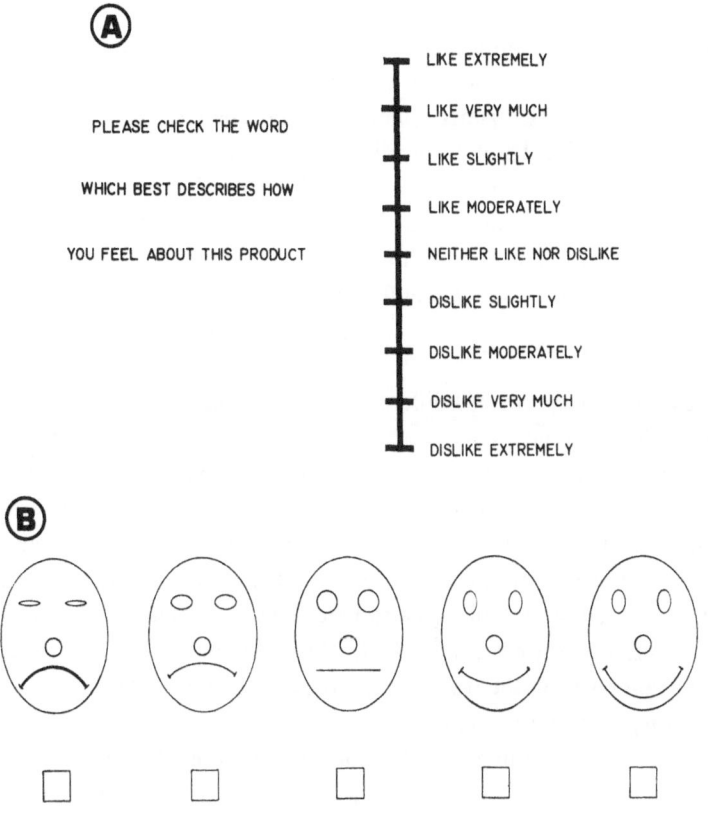

PLEASE CHECK THE WORD

WHICH BEST DESCRIBES HOW

YOU FEEL ABOUT THIS PRODUCT

LIKE EXTREMELY

LIKE VERY MUCH

LIKE SLIGHTLY

LIKE MODERATELY

NEITHER LIKE NOR DISLIKE

DISLIKE SLIGHTLY

DISLIKE MODERATELY

DISLIKE VERY MUCH

DISLIKE EXTREMELY

PLEASE CHECK THE BOX UNDER THE FIGURE WHICH BEST

DESCRIBES HOW YOU FEEL ABOUT THIS PRODUCT.

Fig. 1.   Hedonic scales. (a) Verbal hedonic scale developed by Peryam and Pilgrim (1957). (b) Facial hedonic scale developed by Continental Can Company, Inc. (Ellis, 1966).

6. Discriminatory Tests

Is there a difference between samples?

There are 2 types of discriminatory tests: threshold tests and difference tests.

6.1. THRESHOLD TESTS

Absolute sensitivity or absolute threshold is the physical intensity of a given parameter at which it can just be perceived by a given individual. It is a concept generally applied to flavor and odor where the level of a specific additive can be controlled and evaluated. Harper (1968) comments that the concept of 'absolute sensitivity seems irrelevant to considerations of texture and consistency. The limiting factor is not zero but is more likely to be set by the properties of the skin and muscle and joint system'. No reports of absolute threshold studies of texture attributes could be found.

The difference threshold is the smallest *change* in the intensity of an attribute which a given individual can detect as being different. It is measured in jnd (just noticeable difference) or jnnd (just not noticeable difference) units. The difference threshold can be determined by using triangle, duo-trio, paired comparison, or any of the other standard difference tests (see general methodology references previously mentioned). In a study of the relation between mealiness in baked potatoes and specific gravity, Murphy *et al.* (1967) used triangle tests to establish the minimum detectable difference or jnd. They found that the detection threshold for mealiness was a specific gravity difference of 0.004 units. The panels showed increased precision when differences were 0.006 and were unable to distinguish differences of 0.002.

6.2. DIFFERENCE TESTS

Difference tests determine only if a difference exists between samples, not the magnitude or direction of the difference.

Difference tests do not require as many panelists as do preference/acceptance tests. It is very important that the panelists be of at least normal sensitivity because often, if no differences are found, the samples are assumed to be identical and no further testing is done. Discrimination is based on all of the characteristics of the samples so care must be exercised in controlling extraneous factors such as color, odor, and flavor when texture differences are being studied. The most commonly used difference tests are the paired comparison, the triangle, and the duo-trio tests. Details of their use are given in texts on general panel methods.

According to the Committee on Sensory Evaluation of the Institute of Food Technologists, discriminatory tests are usually performed by 3–10 trained persons or by 80+ nontrained persons (ASTM, 1968a). Persons of low sensitivity should not be included just to achieve a predetermined panel size. Procedures for selecting panelists are described in the ASTM *Manual on Sensory Testing Methods*. Panelists should be of at least normal sensitivity in their ability to discriminate and reproduce their results as determined by threshold, difference, or other tests. They must be trained in the method to be used and are often trained to discriminate very fine differences in specific attributes. It is suggested (ASTM, 1968) that training persons to recognize certain features in a set of reference standards may reduce their reliance on preferences and help them develop more objective judgments.

7. Descriptive Tests

How much difference is there between samples?
What is the difference between samples?

It is seldom of interest to know only if one sample differs from another. Usually, it is desirable to know the amount or direction of difference. According to Scott Blair (1969):

Panels of subjects can, in practice, provide reproducible and useful information in assessing the textural properties of materials in the hands and in the mouth. Not only can they give judgments of the type 'this is firmer than that', but they can produce scales of sensations with some degree of reproducibility; i.e., 'sample B lies half way between samples A and C in firmness'.

Laboratory panel tests which indicate the relative amounts or the direction of differences among samples are *descriptive*. Unfortunately, many supposedly descriptive tests fail to achieve the desired objectivity because of misunderstanding of the terms by the researcher or by the panelists. Many use a personal opinion of the quality based on preference instead of a relatively objective naming of the intensity of the attribute. For example, instead of using 'much too hard' to describe a melon sample, 'very hard' would be preferable because some people prefer hard melons but are fully aware that the melons *are* hard. There are several kinds of descriptive tests: ranking, rating or scoring, and profiling.

7.1. RANKING

In ranking, several samples are presented and the panelist is asked to place them in the order of increasing intensity of a particular attribute, e.g., increasing toughness of four meat samples. There is no need for extensive training after the panelists understand the meaning of the attribute being evaluated. Standards and knowledge of the possible range of the attribute are unnecessary since the technique is a direct comparison of samples presented. Direction of differences is clearly shown by the ranking, but the only indication of the degree of difference or similarity is the frequency with which the order of adjacent samples is confused.

Kuhn *et al.* (1959) had trained panelists rank boiled potato samples from 'soggy' to 'mealy'. A large number of samples were ranked in sets of three in a balanced incomplete block replicated five times. The sum of all sensory scores was termed a 'texture score'. Differences in total texture scores were due primarily to variety. A significant correlation was obtained between texture scores and specific gravity. 'Soggy' and 'waxy' are terms commonly used in the potato industry and are understood to be on a continuum by the panelists. Caution must be exercised in the use of such terms and examples should be presented in training sessions so there can be no confusion of the meanings of trade terms.

7.2. RATING OR SCORING

In rating or scoring, the relative intensity or amount of a property is indicated by a mark on a rating scale or by a numerical score. When panelists are asked to score the intensity of a characteristic on a numerical scale such as 0–100 or the 0 to 3 scale described in the General Foods Texture Profile, the method is generally referred to as 'scoring'. The ratio-scoring system used by Stevens and Guirao (1964) is seldom used because of the complexity of analysis. When a graphic scale is used, the method is usually called 'rating'. Rating scales appear more often in the literature than scores. They usually consist of 5 to 10 points. It is essential to use objective terms, not preference terms in defining the scale points. The trained panelists used for descriptive tests cannot be considered typical consumers; therefore, hedonic scales should not be used. The following quotation from Jellinek (1964) emphasizes this requirement:

What might be 'good' for one person might be only 'fair' for another. No true quality measurements can be expected with such a score sheet because it reflects a mixture of quality rating and hedonic evaluation. Therefore, the scores have to be anchored and defined, when a true quality rating is to be expected.

Use of rating scales or scoring requires a more highly trained panel than does ranking. They must be familiar with the range of the attribute under study. The samples are usually presented and rated separately with some time elapsing between samples so there is no direct comparison. One or several attributes may be evaluated at once. Usually panelists are given rating scales with some or all points anchored verbally or by presenting standards. Numbers are assigned to the points for statistical treatment. Rating scales may be unstructured or structured, as illustrated in Figure 2, or may be intermediate between these.

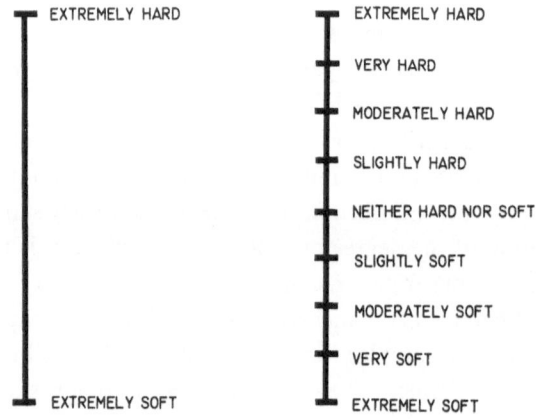

Fig. 2.   Unstructured and fully structured rating scales.

An unstructured scale may be considered to be a line having no discrete points marked, with only the ends identified as extreme intensities of the parameter. An example of such a scale would be a 100 mm line with the ends identified as 'softest' and 'hardest'. The panelist is asked to mark the line at the position he thinks represents the relative intensity (hardness or softness) of the sample. Data are taken as the distance from one end to the panelist's mark. Stinson and Huck (1969) report using unstructured scales for trained panel evaluation of pastry tenderness for correlation with several instrument measurements. No other reports of the use of such unstructured scales in texture studies were found. Unstructured scales are useful in determining the relation between apparent and actual intensity of some textural properties for weighting the apparent intensity scores. A fully structured scale would have all points named and the panelists would be required to use only those points. For intermediate forms, panelists may be instructed to use only the points marked or to mark anywhere on the line using the points as guidelines.

Scale points are generally assumed to divide a physical continuum into equal segments but, because of the nature of human sensitivity, the psychological increments may not represent equal physical increments. For example, Stevens and Guirao (1964) found that perceived viscosity increased as the 0.5 power of actual viscosity and that apparent hardness of rubber balls squeezed between the fingers increased as the 0.8 power of actual hardness. Nonlinear relations between perceived and actual intensities of various physical parameters may be the cause of poor correlations between sensory and instrumental measurements in many cases. Much more research is needed in this field.

If the property is unidirectional, a scale may be from none to extreme, such as 0 = not mealy, to 5 = extremely mealy. In cases where the property can be represented by antonyms, the scale may be from negative to positive with a midpoint at 0 or from a low to a high positive number. The number of points on the scale should be related to the number of intervals distinguishable by the panelists. Generally, discrimination increases with increased scale length from 5 to 9 points (ASTM, 1968a).

Drake (1965b) found no significant differences in mean scores for meat toughness/ tenderness among judges who only rated toughness/tenderness and judges who both rated and tape-recorded the sounds they made during mastication. The panels used a '9-point unstructured scale' with 1 = tough through 9 = tender. Kapsalis et al. (1970) had 4 trained judges rating several factors of meat tenderness on a "9-point unstructured scale with the end points being 'extremely difficult to bite through' and 'extremely easy to bite through'." What is apparently meant in both examples are scales with 9 discrete points but with only the end points named.

An unusual rating ballot, combining two parameters of fish quality, was used by Cowie and Little (1966). The ballot had a 5-unit horizontal scale from tough to soft and a 5-unit vertical scale from wet to dry. These created 25 boxes to indicate simultaneously the moistness and the tenderness. The authors point out that this type of a scale makes it easy to see the properties characterizing fish texture which would be masked by a simple 0–5 overall texture scoring. The same effect would, however, be achieved by using two conventional 0–5 scales, one for wetness and one for toughness.

Bockian et al. (1958) compared sensory ratings of beef tenderness with objective measurements of the work required to grind similar samples. Eleven judges were selected after preliminary panel sessions in which four meat samples were evaluated then discussed to reach common judgments of the degree of tenderness. Samples were rated on a 7-point scale from very tough to very tender. During the two years of the test, correlation coefficients of $r = -0.59$ and $r = -0.60$ were obtained between the sensory and objective measurements. These are significant at 1% but indicate that only about one-third of the variations in tenderness scores can be explained by the work needed to grind the meat and vice versa. Fielder et al. (1963) used a six-member panel to score flavor, juiciness, and tenderness of beef on 9-point intensity scales. They found high correlations between tenderness scores and shear values of the cooked meat for some cuts of meat and some cooking methods. The poor relationships for other cuts and cooking methods may indicate that different parameters are being measured by the panel and the instrument and that shear is not the important parameter in those cases where low correlations were obtained. These are cases where texture profiling could be applied to evaluate the differences between meats giving high and low correlations with shear values. From those results, it might be possible to select better parameters for the routine sensory tests.

There are several errors in usage evident in the following example. In a test of the effects of gamma radiation on shelf life of fruits and vegetables, Truelson (1963) selected eleven tasters on the basis of their sensitivity to four basic tastes. They scored for both flavor and texture.

'Ranks ranged from 0 to 6 (5 = good, 3 = acceptable, 1 = poor) for both taste and texture.... The decisions of the judges were checked throughout the experimental period. If the markings of any judge proved to be inconsistent or based on different levels on different days, the judge was replaced...' Ranks are the order of the samples based on increasing or decreasing intensities of a particular characteristic; the panel used hedonic ratings or scores. The terms are appropriate for consumer tests but selection of panelists by other than random means disqualifies them as typical consumers and a panel size of eleven is too small for a consumer preference test. Scales for flavor and texture intensities (similar to the apple flavor and hardness scales in Figure 3) might be more appropriate. The replacement of panel members in a small panel makes comparisons of results at different sessions difficult. In some circumstances such comparisons are of little concern but in a shelf-life study they are usually important

7.3. MULTIFACTOR RATING SCALES

Texture is not one single property but is a composite of several parameters. To provide a relatively complete word-picture of a food, a multifacet approach is required. The several components of the quality of a food must be identified and evaluated.

During extensive studies of meat quality, a specialized profile-type method was devised by Cover *et al.* (1962) for evaluating factors involved in meat tenderness. Since tenderness is related to the characteristics of muscle fibers and connective tissue, separation for sensory evaluation seemed logical. Tenderness was separated into six component factors for panel judgment: overall softness to tongue and cheek, softness to tooth pressure, ease of fragmentation of muscle fibers across the grain, mealiness upon chewing, apparent adhesion between fibers, and the amount and hardness of connective tissue. These factors were placed on 9-point rating scales for panel use. Cover *et al.* (1962) reported that 'use of scores for the connective-tissue component of tenderness has increased the precision of this phase of the attack on the multiple causes of tenderness.' This is a multifactor rating method which was a forerunner of the specialized 'profile' method. The General Foods Texture Profile utilizes rating scales for the principal factors but relies heavily on verbal descriptions as well.

Figure 3 is a trial ballot used by several researchers in the Agricultural Marketing Research Institute of the U.S. Department of Agriculture investigating methods of assessing apple quality. This is another multifactor rating method. Texture is rated on 9-point scales for overall texture, toughness, and hardness. The terms of the scale headed 'texture' are accepted terms in the apple industry and are used in USDA *Grades and Standards*. The inclusion of toughness and hardness scales, suggested by Finney (1969b), is an attempt to find the components of texture for correlation with instrumental measurements. In using this ballot, the author and associates have found that panelists, including several experienced pomologists, do not concur on the mean-

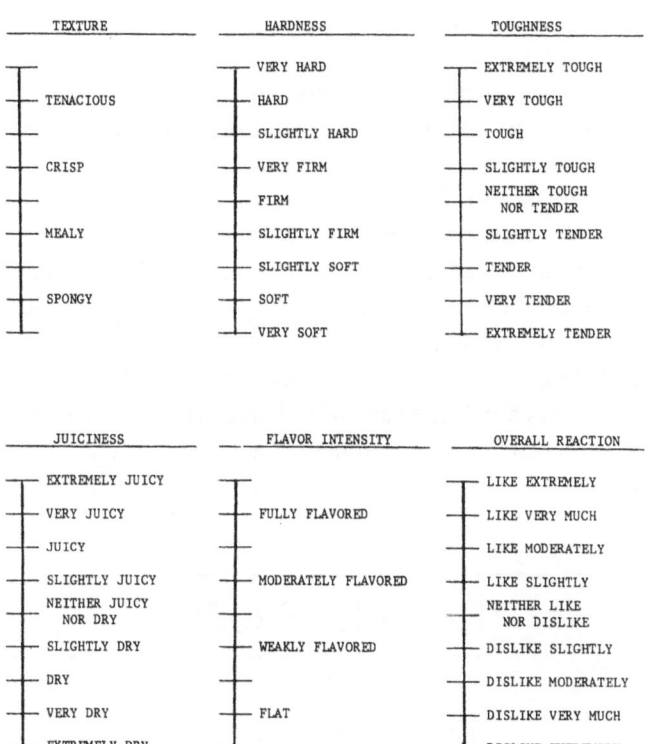

Fig. 3.   Trial ballot for apples (Quality Evaluation Investigations, Market Quality Research Division, U.S. Department of Agriculture).

ings of the terms in the *texture* scale. Panelists also felt that using only the one general texture scale was inadequate and that the terms did not fall on a continuum. If a similar ballot were constructed on the basis of a texture profile, the terms would have been thoroughly discussed, other terms suggested, and any disagreement settled by use of other commodities as examples. Several additional parameters would probably have been added. It is recognized that five trained panelists are not a consumer panel but the hedonic scale was included as a general indication of acceptability for the researchers. The panelists seem to be more objective in scoring the quality factors if they are provided with a means of expressing their opinions of the fruit.

7.4. TEXTURE PROFILE

In the flavor profile developed by the Arthur D. Little Co. in 1950 (Cairncross and Sjöström, 1950), flavor is described in terms of flavor-note identity, order of perception, relative intensity, amplitude and aftertaste. In 1963, researchers at General Foods Corporation published a method for profiling texture (Brandt *et al.*, 1963, and Szczesniak *et al.*, 1963a). The panel procedures described by Brandt *et al.* (1963) involve the evaluation of the mechanical, geometrical, fat and moisture properties of foods. Unlike flavor notes whose order cannot be predicted, texture characteristics are perceived in an ordered sequence. In the texture profile, analysis of the properties is done at each of the three stages of ingestion: initial, masticatory, and residual. Figure 4 summarizes the General Foods Texture Profile parameters.

The initial stage of ingestion, the first 1–5 chews depending on the product, involves the mechanical properties of hardness, brittleness, and viscosity. The masticatory stage, extending to the time the food is ready for swallowing, may involve gumminess, chewiness, and adhesiveness, as well as hardness, brittleness, and viscosity. At the residual stage, the changes made during mastication and the condition of the residue are evaluated. This may include notation of any of the previous mechanical properties but is especially concerned with the type and rate of breakdown and the general mouthfeel. At all three stages, the geometrical and other properties are noted.

Mechanical properties are defined as those characteristics related to the reaction of the food to stress. These include hardness, fracturability, chewiness, gumminess, springiness, adhesiveness, and viscosity as explained in Table I. The definitions are

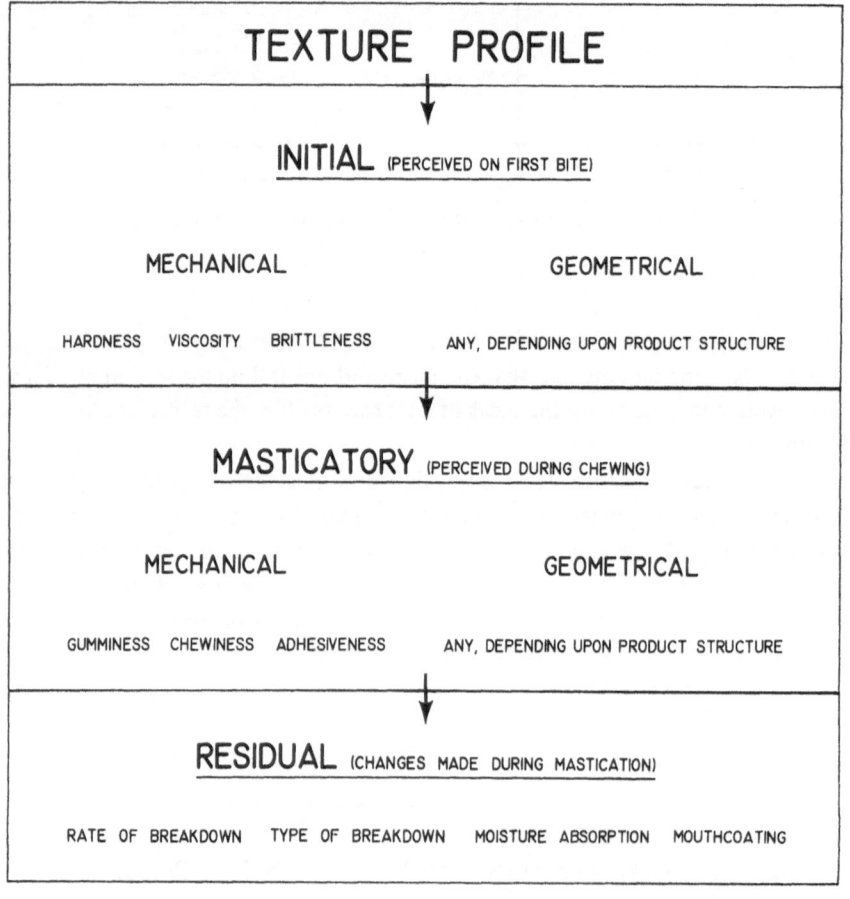

Fig. 4.   Procedure for evaluating texture. (Adapted from Brandt *et al.*, 1963).

TABLE I

Mechanical characteristics from the General Foods Corporation Texture Profile: Sensory definitions and common descriptive terms. (Adapted from General Foods, 1970.)

Primary Properties

Hardness – The force required to compress a substance between the molar teeth (for solids) or between the tongue and palate (for semi-solids) to a given deformation or to penetration. *Soft, firm, hard.*

Cohesiveness – The extent to which a material can be deformed before it ruptures.

Viscosity – The force required to draw (slurp) a liquid from a spoon over the tongue. *Thin, watery, thick.*

Springiness – The amount of recovery from a deforming force; the rate at which a deformed material returns to its undeformed condition after the deforming force is removed. *Elastic.*

Adhesiveness – The force required to remove material that adheres to the mouth (generally the palate) during the normal eating process. *Sticky, tacky, gooey.*

Secondary Properties

Fracturability – The force with which a sample crumbles, cracks, or shatters; the horizontal force with which the fragments move away from the point where vertical force is applied. Fracturability is the result of a high degree of hardness and low degree of cohesiveness. *Crumbly, crunchy, brittle.*

Chewiness – The length of time or the number of chews required to masticate a solid food to a state ready for swallowing. Chewiness is a product of hardness, cohesiveness, and springiness. *Tender, chewy, tough.*

Gumminess – A denseness that persists throughout mastication; the energy required to disintegrate a semi-solid food to a state ready for swallowing. Gumminess is a product of a low degree hardness, and a high degree of cohesiveness. Scale is based on forty to sixty percent flour and water pastes in five intervals. *Short, mealy, pasty, gummy.*

based on sensory techniques instead of on instrumental tests in order that panelists can better relate them to their own sensations. Geometrical properties are related to the arrangement of the physical constituents within the food. Other properties are fat and moisture content and rate of release.

Standardized rating scales were developed for evaluating the sensory characteristics of mechanical properties in conjunction with instrumental texture measurements as described by Szczesniak *et al.* (1963a). The major texture parameters were determined for a wide variety of foods. Anchored rating scales were set up with each intensity increment represented by a specific example, designated by brand, size, and temperature of serving. These examples served as reference standards. Table II lists the generalized standards for the main mechanical parameters.

Since the standardized rating scales were intended to cover the full intensity range of each property, they are very broad. During training programs, the standardized scales provide a frame of reference for teaching the descriptive and quantitative aspects of food texture attributes in general. Since variations of a certain food type could fall within one or two points on a given standardized scale yet be significantly different from each other, General Foods panels have found it convenient to use relative rating scales to describe each parameter in terms of typical intensities for that food, using the flavor profile scoring system. If a normally present parameter is not detectable in one sample, it is rated 0; if barely detectable, threshold)(; if strong, 3. No points can be used between 0 and threshold. Between)(and 3, a 13-point scale is

TABLE II

Standardized rating scales for mechanical properties with generalized standards[a, b]
(intensity of parameter increases downward)

Hardness	Fracturability	Chewiness
Cream cheese	Corn muffin	Rye bread
Velveeta cheese	Egg Jumbos	Frankfurter
Frankfurters	Graham crackers	Large gum drops
Cheddar cheese	Melba toast	Well-done round steak
Giant stuffed olives	Bordeaux cookies	Nut Chews
Cocktail peanuts	Ginger Snaps	Tootsie Rolls
Shelled almonds	Treacle Brittle	
Rock candy		

Adhesiveness	Viscosity
Hydrogenated shortening	Water
Cheese Whiz	Light cream
Cream cheese	Heavy cream
Marshmallow topping	Evaporated milk
Peanut butter	Maple syrup
	Chocolate syrup
	Cool 'n Creamy pudding
	Condensed milk

[a] Adapted from Szczesniak et al. (1963a) and General Foods (1970).
[b] For specific brands, dimensions, and serving conditions of standards, refer to Szczesniak et al. (1963a) or General Foods (1970).

TABLE III

Geometrical properties[a]

Related to particle size and shape		Related to particle shape and orientation	
Property	Example	Property	Example
powdery	confectioners sugar	flaky	flaky pastry
chalky	tooth powder	fibrous	breast of chicken
grainy	cooked Cream of Wheat	pulpy	orange sections
gritty	pears	cellular	apples, cake
coarse	cooked oatmeal	aerated	whipped cream
lumpy	cottage cheese	puffy	puffed rice
beady	cooked tapioca	crystalline	granulated sugar

[a] Adapted from General Foods (1970).

used with increments such as 1, 1–2, 1–2, 1–2, and 2. The 1–2 is read '1 to 2 on 1' or '1–2 nearer 1' meaning that the intensity is half way between 1 and 1–2, or is slightly greater than 1. These ratings are used to indicate the intensities of the mechanical, geometrical, and other properties.

Geometrical properties may be related to particle size and shape, such as powdery or beady, or to particle shape and orientation, as flaky or fibrous. Geometrical terms and examples appear in Table III. The words in the 'particle size and shape' list in Table III imply increasing particle size.

Moisture and fat content, time, and manner of release are also described.

In a complete texture profile, any characteristic perceived is noted. It is not practical or necessary to develop standardized scales for all of the textural characteristics found in all food products. Frequently, special properties, such as sliminess in gum solutions, are found to be important in a particular food class but are not found in most foods. Other special characteristics are uniformity, multiple phases, greasy mouthcoating, rate of breakdown, etc.

The key to a successful use of the texture profile method is the recognition of its flexibility. The method is adaptable to different products and different objectives. When the major texture differences among samples of a product type are other than the mechanical properties exemplified on the standard scales, appropriate procedures and terminology are developed during preliminary testing sessions. Ballots listing the expected parameters are used in routine sessions, but provision is made for comments and for adding new parameters as the panel finds additional ones of importance.

Immediately after individual evaluation of the samples, the panel leader compiles the scores. Any major disagreements are resolved by discussion and reexamination by the panel and referral to reference standards if necessary. The leader then writes a composite report. In this way, continual training and development of the method occur since the discussions bring out problems and suggestions.

A valuable use of the profile method would be to establish and define the important parameters for evaluating quality differences among samples of a food. These could then be put on simplified rating scales for use by less highly trained panelists in routine quality testing.

8. Summary

The methods used for sensory evaluation of flavor or nearly any other sensory factor. can be adapted for evaluating texture. The correct uses and limitations of each panel method must be considered in selecting the method and in interpreting the results of sensory testing. Before establishing a sensory testing program, consult the sensory evaluation manuals and if possible, before selecting a method, experiment with some preliminary panels to become familiar with the mechanics of each of the methods which might satisfy the requirements of your program. There are right and wrong ways to organize sensory panels but there is no 'only way'. Often a sequence of different tests is more efficient than using only one kind. For example, one might use difference tests to establish that samples from several treatments are different. One might then have profile panels examine samples to describe how they differ, then use those parameters in rating tests for routine comparison of many samples.

A panel should not be asked to perform too many tasks at once. Limitations are imposed by the test method and by the abilities of the panelists. The number of questions to be answered must be limited and the number of samples to be evaluated must be restricted. Remember, these are people, not machines, and they will tire and loose accuracy if overworked.

Guide to using sensory panels

Requirement	Type of panel		
	Preference acceptance	Discriminative	Descriptive
To determine consumer likes and dislikes	+	−	−
To create a new product with no prototype	+	−	+
To match a prototype or standard	−	+	+
To improve a product	+	+	+
To evaluate effects of packaging, storage, change in process, change in ingredient	−	+	+
To define specifications	−	−	+
Quality control	−	+	+
To measure sensitivity	−	+	−

Panel method	Function	Recommended panel size
Preference/Acceptance	*Evaluate consumer opinions*	
Hedonic (verbal or facial)	rate samples individually for degree of acceptability	80–120 nontrained preferred, at least 30 for rough product screening
FACT (Food Action)	rate samples individually on an action scale	
Paired comparison (hedonic)	select preferred sample from two samples	
Ranking (hedonic)	compare several samples for degree of acceptability	
Discriminative	*Determine whether differences exist among samples*	
Threshold	determine human sensitivity to specific characteristics	nontrained, as needed
Differences	detect differences among samples	
Degree of difference or scalar difference	rate degree of difference between samples and standard	5–10 trained or 8–25 semitrained
Paired comparison	detect difference between two samples	
Triangle	detect difference between two samples by separating odd sample from two identical samples	5–10 trained, 8–25 semitrained, 80 + nontrained
Duo-trio	detect difference between two samples by matching one with a standard	
Other variations		
Descriptive	*Evaluate the product in terms of specific characteristics*	
Ranking	compare several samples for intensity of specified characteristics	5–10 trained, 8–25 semitrained, 80 + nontrained
Rating	evaluate individual samples for intensity of specific characteristics	5–10 trained
Profiling	verbal description of product in combination with rating scales for some characteristics	4–6 highly trained

CHAPTER IV

ELEMENTARY CONCEPTS OF RHEOLOGY RELEVANT TO FOOD TEXTURE STUDIES

ESSEX E. FINNEY, Jr.

1. Introduction

Rheology is a branch of physics which deals with the deformation and flow of materials, both solids and fluids (Reiner, 1960). In the case of food materials, their rheological (deformation and flow) behavior is directly associated with their textural qualities. Consumers, for example, estimate fruit firmness on the basis of the deformation resulting from physical pressure applied by the hand and fingers. The toughness or tenderness of meat is subjectively evaluated in terms of the effort required for the teeth to penetrate and masticate the flesh tissues.

Instrumental (objective) tests for texture also rely upon the deformation and flow characteristics of the food material (Finney, 1969a). The attribute of toughness, for example, has been related to the maximum force or to the energy required to shear through a food sample (Backinger, et al. 1957). Firmness is often specified in terms of the force required to achieve a given deformation of a body (Finney, 1969b, Szczesniak and Bourne, 1969). A viscosity-type test frequently is used to evaluate the texture or rather consistency (resistance to deformation: Schmidt and Marlies, 1948) of soft foods such as puddings, sauces, purees, and similar fluids. For the aforementioned reasons, concepts from rheology traditionally have been considered important from a theoretical as well as experimental viewpoint during studies of food texture.

Foods, however, are complex materials structurally and rheologically. In many instances they consist of mixtures of solid as well as fluid structural components, e.g., solid cellwall material, water and colloidal liquids, and intercellular gases. Many foodstuffs are neither homogeneous nor isotropic but have properties that vary from one point to another within their mass. In spite of these complicating factors, investigators report that many foods do behave in a predictable manner and concepts from elasticity, plasticity, and viscosity theories can be used to interpret their responses to applied force. These rheological concepts should therefore be given adequate treatment in investigations of the textural properties of foods.

2. Preliminary Considerations

2.1. SOLIDS AND LIQUIDS

Materials consist of particles which are bound together in various ways. Of the three commonly recognized states of matter, only two – solids and liquids – are of primary

A. Kramer and A. S. Szczesniak (eds.), Texture Measurements of Foods, 33–51. All Rights Reserved.
Copyright © 1973 by D. Reidel Publishing Company, Dordrecht-Holland.

interest in studies of food texture. *Solids* are capable of retaining a definite size and shape and of resisting (up to a certain limit) forces which tend to deform. *Liquids*, on the other hand, are substances which are incapable of sustaining a shear stress. They flow readily and tend to assume the shape of the confining vessel.

Many foods are neither perfectly solid nor perfectly liquid, but exist in a state of aggregation which is often referred to as '*semi-solid*,' i.e., a state intermediate between solids and liquids. Elder and Smith (1969) list gelatin desserts, jellies, gum confections, peanut butter, puddings, and the like in this category. Such foods may also be considered as soft solids, i.e., they do not flow as liquids but they do exhibit gross distortion under moderate stress, such as the force of gravity. The boundary between solids and semi-solids is vague and arbitrary. Ferry (1961), however, has suggested that materials having shear modulus values less than 10^9 dyne cm^{-2} may be classified as 'soft' viscoelastic solids. This, nevertheless, is still somewhat arbitrary since the elastic modulus of a material, especially a polymeric system, may depend to a large extent upon test conditions such as the rate of uniaxial deformation or, in the case of sinusoidal dynamic testing, the oscillatory frequency.

Finally, there is another category referred to by Matz (1962) as 'compound foods'. These are predominantly liquid in behavior but include perceptible amounts of non-liquid particles, e.g., soups, purees, sauces, and foams. Sherman (1971) has pointed out that these are essentially 'dispersed systems' and the term 'compound foods' is not a very appropriate designation. Be that as it may, Elder and Smith (1969) emphasize that rheological principles apply to all of these varied and complex food agglomerations since rheology is broadly defined as 'the deformation and flow of matter.' Complex structures, however, do increase the difficulty of relating theory to experimental observations.

2.2. FORCE AND DEFORMATION

It is convenient to consider concepts of force and deformation simultaneously. This is because the two phenomena coexist within a given mass, i.e., deformations within a body can be achieved only through the application of force, and vice versa.

Force is any influence which causes a change in the state of motion of a material or which maintains an elastic material in a deformed configuration. The motion may consist of a translation and/or rotation which involves no change in the relative position of particles within the mass (rigid body motion), and/or a *deformation* wherein a change occurs in the relative positions of particles within the body. Deformation is the term generally used in referring to solids. Flow is a time-dependent form of deformation, which is used more commonly in referring to the motion of fluids. Deformations, in general, may be conservative or dissipative with respect to mechanical energy within the deformed material. Classical examples of these two types of behavior are elastic (recoverable) deformations and viscous (non-recoverable) deformations (flow), respectively. Intermediate between the two extremes is a type of mechanical deformation which is termed 'viscoelastic,' i.e., it is neither entirely conservative nor entirely dissipative, but combines these two effects.

2.3. STRESS AND STRAIN

As a result of the application of force, a body experiences deformation. In addition to surface stress and strain, the body also develops internal stresses and strains. Standard definitions of the various types of stress and strain configurations developed within materials are presented in detail within ASTM Designation: E6–66 (ASTM, 1968b). Restatement of each of the definitions is neither necessary nor feasible. Certain of the terms, however, such as stress, strain, tensile strength, compressive strength, and shear strength, are frequently found in the literature on food texture. Hence, a brief discussion of these terms may be worthwhile.

Stress is the intensity, at a given point in or on the surface of a body, of the components of the force that acts on a given plane through the given point. It is expressed in terms of force per unit area. We speak of tensile, compressive (σ), and shearing stresses (τ), depending upon whether the stress components are directed away from, toward, or tangentially to the plane on which the force components act. Of the two first mentioned, tension is considered positive.

Strain is the mathematical expression of a change in the size or shape of a body with reference to its original size or shape. For example, in a tensile test, strain $\varepsilon = \Delta L/L$, where L = length. Strain is a nondimensional parameter (ratio or percent), but it is frequently expressed in inches per inch or centimeters per centimeter. Just as there are three sorts of stress, there are also three sorts of strain: tensile, compressive, and shear.

The *tensile* or *compressive* strength of a material is the maximum tensile or compressive stress which a material is capable of sustaining before rupturing. It is calculated from the maximum load during a tensile or compression test and the cross-sectional area of the specimen. Similarly *shear strength* is the maximum shear stress which a material can withstand. It is calculated from the maximum load during a shear (e.g., torsion) test and the geometry of the specimen.

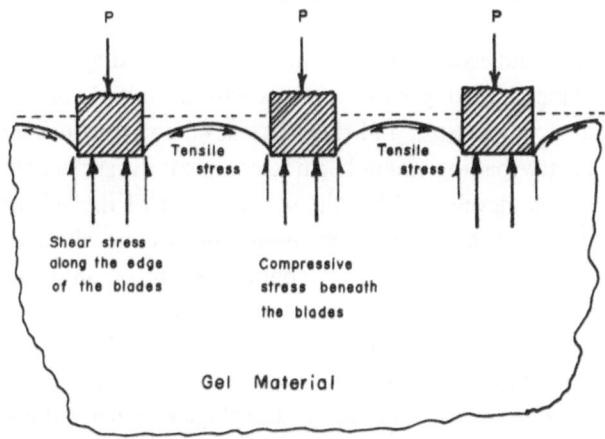

Fig. 1. A hypothetical case illustrating various states of stress possible in a gel loaded with a multiple blade shear press cell.

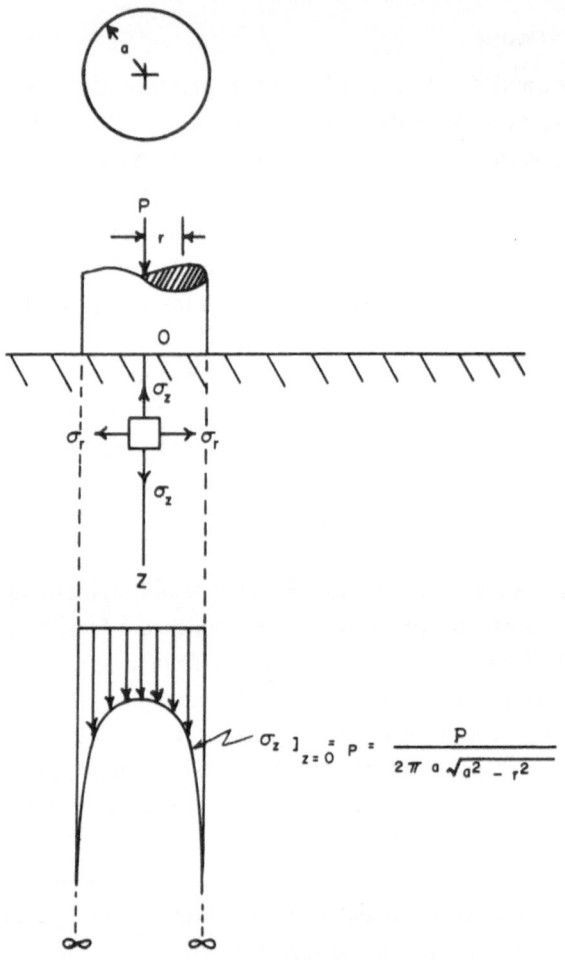

Fig. 2. Theoretical stress distribution under a rigid die acting against a semi-infinite elastic body.

In many instances, and especially in testing foodstuffs, a complex state of stress may exist within the sample. During measurements with a multiple-blade shear press, as an example, tensile, compressive, and shear stresses may exist during early phases of loading of a gel. As the loading blades begin to compress the surface (Figure 1), tensile stresses will, as a rule, develop within the surface layer of the gel between adjacent blades. Compressive stresses will exist below the blades and shear stresses develop, especially along the edges. Similar stress conditions can be anticipated when rigid cylindrical dies (punches, Figure 2) are used to puncture or only load food materials (cf., Timoshenko and Goodier, 1970; Finney and Hall, 1964).

The aforementioned definitions of stress and strain are applicable to all materials regardless of their structure. Stress-strain relationships within a material, however, are related to structure and depend upon the manner in which the material responds to force or to *an imposed deformation*. Theories relating stress to strain within materials may be broadly considered under the headings of 'elasticity,' 'viscosity,' and 'plasticity,' with various combinations of the three to account for overlapping behaviors such as viscoelastic or elastoplastic behavior (Ferry, 1961; Bland, 1960; Reiner, 1960).

3. Types of Material Behavior

From a rheological viewpoint, three types of deformation are generally recognized, i.e., elastic, plastic, and viscous. Even though few materials are either perfectly elastic, plastic, or viscous, theories from elasticity, plasticity, and viscosity have importance in many studies of the rheological behavior of materials. Such concepts also have significance to many of the investigations of textural properties of foods.

3.1. ELASTICITY

A body is perfectly elastic if deformation or strain occurs instantaneously with the application of stress and this deformation completely and instantaneously disappears when the stress is removed (Fitzgerald, 1961). In addition, it is generally assumed that there is a one-to-one relationship between the state of stress and the state of strain in an elastic body; hence, all time-dependent effects are excluded.

3.1.1. *The Ideal Elastic Solid*

For small strains, certain bodies under axial stress exhibit ideal elasticity and stress is directly proportional to strain $\varepsilon = \Delta L / L$, where $L =$ length. The proportionality constant has been called 'Young's modulus' or the 'modulus of elasticity,' E. The relationship between stress and strain, therefore takes the form:

$$\sigma = E \cdot \varepsilon \text{ (uniaxial stress)}.$$

A solid body which obeys this equation is often referred to as a 'Hookean' body in honor of Robert Hooke, who suggested this relationship in 1676. This simple equation is a special case of the more generalized constitutive equations of elasticity which are discussed in greater detail by Sokolnikoff (1956) and Malvern (1962).

In addition to the equation for the uniaxial loading condition, analogous expressions may be written for torsion loading (shear stress vs shear strain) and for hydrostatic or bulk loading (volumetric stress vs volumetric strain). In the latter two cases, the proportionality constants are the shear modulus G, and the bulk modulus K, respectively.

If in addition to satisfying small-strain loading conditions, the materials also are homogeneous and isotropic, then the elastic constants, E, K, and G are interrelated by the expressions:

$$K = E/3 (1 - 2\mu)$$

$$G = E/2 (1 + \mu),$$

where μ is Poisson's ratio. Poisson's ratio is the ratio of the percent lateral expansion to the percent uniaxial compression when, for example, a cylindrical specimen is subjected to a compressive force (Chappel and Hamann, 1968).

Typically, the elastic properties of materials are measured using cylindrical specimens of known dimensions. Such specimens are subjected to uniaxial tension (elongation) or compression (contraction). It is also possible, however, to calculate the elastic properties of materials having other shapes and loading configurations. Many of the mechanical tests and equations appropriate for calculating the apparent elastic properties of solid food materials have been tabulated by Mohsenin (1970a, b).

References in the food literature to 'elastic modulus' or 'modulus of elasticity' refer to Young's modulus of elasticity or the ratio of stress to strain during uniaxial loading. A few investigators, however, also have investigated the shear elastic modulus (Hibberd and Wallace, 1966; Hibberd et al., 1966; Finney, 1967) and the bulk modulus of elasticity of solid food materials under hydrostatic pressure (Morrow and Mohsenin, 1966; Finney and Hall, 1967; White and Mohsenin, 1967; Sharma and Mohsenin, 1970).

3.1.2. Elastic vs Textural Characteristics of Foods

Much attention has been devoted to studies of the modulus of elasticity of foodstuffs and its relationship to structure, texture, and overall quality. Cornford *et al.* (1964), for example, measured the elastic modulus of bread in an effort to evaluate the relationship between changes in crystallinity of starch gel and bread staling. They noted that bread staling is accelerated by a reduction in temperature, down to the freezing point of bread. Likewise, the rate of crystallization in a high-polymer system becomes progressively faster as the temperature falls below the melting point of the crystalline phase. Changes in the modulus of elasticity of the bread crumb, under specified conditions of strain and time, were used to follow the course of staling. The elastic modulus increased with age of the bread (time after baking), increased as the storage temperature decreased, and increased for bread baked without fat as opposed to that prepared with fat. Reported values for elastic modulus generally ranged from 25 to 300×10^3 dynes cm^{-2}, with the higher values being associated with the aged stale bread.

As a result of a similar study, Elton (1969) reported a close correlation between the modulus of elasticity of bread and sensory ratings of freshness. Panel ratings ranged from 'quite fresh' for a crumb modulus of 50×10^3 dynes cm^{-2} to 'barely fresh' for 100×10^3 dynes cm^{-2} to 'quite stale' for a modulus of 200×10^3 dynes cm^{-2}.

Young's modulus of elasticity is often referred to by cereal chemists as a 'modulus of compressibility' (Bice and Geddes, 1949). It should be noted that this use of the term 'compressibility' in cereal chemistry may be quite different from the use normally found in physics or materials science. In physics, compressibility is the reciprocal of the bulk modulus K, or the modulus of elasticity as measured under a uniform hydrostatic stress. Many measurements in cereal chemistry, moreover, are performed under other than small-strain conditions. Caution should be used, therefore, when interpreting and comparing the published results of different researchers.

In a study of biochemical changes associated with tenderization of chicken meat,

DeFremery and Pool (1960) measured the modulus of elasticity of 4–6 cm long strips of muscle approximately 1 cm^2 in cross-section. The elastic modulus, E, in pre-rigor muscle was $1-2 \times 10^3$ gm cm^{-2}. This value increased 10 fold, as the muscle developed a rigid, nonplastic texture and rigor became fully established. The increase in the modulus that occurred during the onset of rigor was not reversed during or following resolution of rigor. Resolution of rigor and the development of tenderness during aging, therefore, did not involve any measurable changes in the elastic modulus of the muscle. The implication is that modulus of elasticity may be related to rigidity and stiffness, but not to meat tenderness.

The elastic properties of fruit and vegetable tissues have been studied in an effort to determine their relationship to texture or to changes in texture during growth, development, ripening, and senescence (Cooper, 1962). Virgin (1955) and Falk *et al.* (1958) developed a resonance technique to study the effects of uptake and loss of moisture upon the rigidity of potato tissues. They reported a linear relationship between modulus of elasticity and the turgor pressure developed within the cells. Somers (1966) reported similar results using cantilever beam theory and measuring the deflection or 'droop' of the free end of beam-shaped potato specimens, deflection being inversely related to the modulus of elasticity or stiffness of the specimen. Somers (1966) reported that very small changes in the fresh weight (moisture) of potato tissue are associated with large changes in the apparent values for E, elastic modulus. Finney and Norris (1967) observed an inverse correlation between elastic modulus and total solids in potatoes. This also suggests a direct correlation between moisture and modulus of elasticity of the tissue.

Even though modulus of elasticity may be directly related to moisture content and turgor within fruits and vegetables, its relationship to texture is not clear. Finney (1969b) suggested that modulus of elasticity should be related to the 'firmness' attribute of texture (Table I), arguing that firmness should be closely allied with those characteristics of materials which physicists call 'stiffness' and 'rigidity.' Whereas this hypothesis may hold true for some materials, results of studies by Szczesniak and Bourne (1969)

TABLE I

Values for the apparent modulus of elasticity of some food products
spanning a wide range of 'firmness.'

		Food Item	Young's modulus	Reference
			(10^5 dyne cm^{-2})	
	more firm	Carrots, raw	2000–4000	Finney (1969)
		Pears, raw	1200–3000	Finney (1967)
		Potatoes, raw	600–1400	Finney and Norris (1967)
FIRMNESS →		Apples, raw	600–1400	Finney (1967)
		Peaches, fresh	200–2000	Finney (1967)
	less firm	Bananas, fresh	80–300	Finney *et al.* (1967)
		Gelatin gel	20	Van Wazer *et al.* (1963)
		Bread	1–3	Conford *et al.* (1964)

indicate that it probably does not apply universally to all food materials. They suggested that in the case of hard or very firm products such as carrots, modulus of elasticity may well correlate with firmness since the sensory judgement of firmness is based upon a bending test. Somers (1966) noted that there is an inverse relationship between the deflection of a beam (degree of bending) and modulus of elasticity. Hence, a hard carrot would bend very little under a given stress, whereas a soft wilted carrot would bend considerably more.

For apples, on the other hand, a puncture test is frequently used to assess firmness objectively. Since there is no basis for assuming an inherent relationship between puncture force and elastic modulus, no sound basis exists for presuming a consistent correlation between modulus of elasticity and the puncture-type firmness measurement in apples. In fact, the small-strain assumptions implicit from theory of elasticity do not apply to the large displacements involved during puncture testing. Hence, any correlations between modulus of elasticity and puncture force should be coincidental. Tests on apples by Finney (1971) indicate that this type of correlation is indeed variable and unpredictable.

3.1.3. *Strength of Food Materials*

In many instances, tensile, compressive, and shear strength measurements are of greater interest than the ratio of stress to strain (E) under small-strain conditions. Tests designed to measure the mechanical strength of foods are inherently destructive, and hence, closely approximate the masticatory process itself which involves the breakdown and destruction of food into a state suitable for swallowing.

Whittenberger (1951) prepared cylindrical specimens from raw apples, evacuated, cooked, and then compressed them using a Delaware jelly-strength tester. He reported good agreement between percent compression under a standard pressure and subjective firmness scores. Tensile strength of the cooked tissues also correlated significantly with compression measurements. He noted, however, that tissues having equal firmness may differ widely in toughness. Tough specimens remained completely coherent even though compressed under relatively high stress. Tough specimens also recovered their original dimensions upon the removal of stress. Tender tissues, on the other hand, completely collapsed under only moderate stress. Since the collapse involved separation of whole, intact cells, Whittenberger indicated that these tension and compression measurements directly reflected the strength of the middle lamella. Such measurements, therefore, might be used to study the role of intercellular cementing substances upon tissue texture.

Others also have used tensile-strength measurements to study structural changes affecting texture of raw and cooked tissues. Personius and Sharpe (1938) used tensile-strength measurements to study the influence of temperature upon the adhesion properties of potato tissue. Huff (1967) measured the tensile strength of potato tissues at rates of strain from 0.0015 to 15 in. in.$^{-1}$/s^{-1} and found wide variations in tensile strength, depending primarily upon the structural zone from which the tissue specimens were taken. Linehan and Hughes (1969) measured the compressive strength of sections of tuber as an index of intercellular adhesion. Intercellular adhesion was one of the important textural characteristics of cooked potatoes.

Halton (1949) reported that tensile strength measures the breadmaking potentialities of flour and the quality of bread that can be achieved under optimum conditions. The tensile strength of the dough, moreover, was a function of the wheat from which the flour was milled and did not appear to be influenced by fermentation, or mechanical manipulation of dough.

In a study of factors affecting the quality of pie doughs and crusts, Miller and Trimbo (1970) showed a good correspondence between tensile strength of dough and toughness of the pie crust. Both increased with progressive decreases in percent shortening in the dough. Matthews and Dawson (1963) found that 79 to 96% of the variation in sensory tenderness scores for pastries was associated with breaking strength as measured objectively on a Bailey shortometer.

Many approaches have been taken toward assessing the strength characteristics of meats. The goal of such studies is to predict tenderness or toughness, the most important attribute of meat texture (Szczesniak and Torgeson, 1965). One of the most popular tests for meat texture is the Warner-Bratzler Shear test (Warner, 1927; Bratzler, 1932). The maximum force required to 'shear' through a core sample of meat with a blade is measured and used as an index of toughness. Such a test is not a pure shear test but involves a combination of shearing, compressive, and tensile stresses.

Tests of the mechanical strength of meat specimens under pure tensile stress, however, have been reported in the literature. Hostetler and Cover (1961) and Cover et al. (1962) for example, measured the amount of extension required to break single fibers of beef muscle. Individual fibers were grasped at two points 5 mm apart and stretched until broken. Extension was measured to the nearest 0.25 mm. The ability of the fibers to extend without breaking (extensibility) was associated with toughness.

Pool (1967) used a tensile test to measure tenacity of connective tissue of muscles from the chicken breast. Uniform cylinders of cooked muscle with the fibers parallel to the plane ends of the cylinder were subjected to a tensile test. The maximum force and work required to tear the fibers apart were measured and recorded. Pool reported that connective tissue tenacity appeared to be a good measure of the connective tissue component of toughness and was independent of the force required to shear across the fibers.

Measurements of the elastic properties and of strength characteristics, therefore, are widely used in testing foods. The appropriateness of any particular measurement, i.e., tension, compression, or shear, however, depends upon the type of food product being tested, and the textural attribute of primary interest.

3.2. VISCOSITY

The viscous behavior of foods is important in many areas of food technology. Ezell (1959), for example, noted that viscosity can become a significant factor during the concentration of citrus juices, especially in the production of high-density concentrates, due to the inefficiency of the operation when the product becomes highly viscous. A very viscous concentrate, moreover, does not reconstitute easily. Hence, viscosity of the product becomes of practical significance to the consumer as well as the processor. Other examples can also be cited, but first let us consider the meaning of the term 'viscosity.'

3.2.1. *The Ideal Viscous Fluid*

An ideal viscous fluid differs from an ideal elastic solid in that it is *rate of strain*, rather than strain, that is proportional to stress in the range of linear behavior. Moreover, the strain or displacement produced at the end of any time period is irrecoverable when the applied stress is removed. Energy is dissipated during the flow process. A characteristic property which determines the rate of flow of any fluid is its *viscosity* which may be broadly defined as its 'internal friction' or 'resistance to flow.' Viscosity, however, also can be defined mathematically. Consider as an ideal case laminar flow of a fluid between two parallel plates one of which is at rest, while the other is moving rectilinearly with a constant velocity '*V*' parallel to itself as shown in Figure 3. The

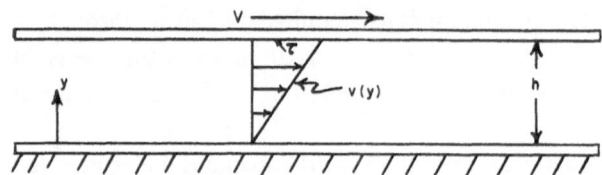

Fig. 3.   Velocity distribution in a viscous fluid moving between two parallel flat plates.

fluid will adhere to both walls (Schlichting, 1968), so that the fluid velocity at the lower plate is zero and at the upper plate is velocity V of the plate. It can be shown that the fluid frictional shearing stress τ is of the form:

$$\tau = \eta \left(-\frac{dV}{dy} \right)$$

where η = coefficient of viscosity, and dv/dy = velocity gradient (shear rate or rate of strain) within the laminar flow. This expression is occasionally referred to as 'Newton's law of internal friction' and can serve as a definition of viscosity. Fluids which behave according to this relationship are called 'Newtonian fluids.'

Viscosity in Newtonian fluids is not only a material property, but it is a material constant that changes only with temperature. Examples of such fluids are water, plant and mineral oils, sugar solutions, and glycerin (Hamm and Kulmbach, 1967).

If the ratio of shear stress to rate of shear is not linear, then the fluid is non-Newtonian. The quantity η no longer represents a simple measure of viscosity, but is a function of the shear rate during testing and under such conditions should be called 'apparent viscosity.'

Typical relationships between shear stress and shear rate for various types of fluids are shown in Figure 4. The following equation may be used to represent certain of these relationships (Charm, 1960; 1962):

$$\tau = b\left(-\frac{dv}{dy}\right)^s + C$$

where b=proportionality factor; s=pseudoplasticity constant; and c=yield stress.

(a) For a Newtonian fluid:

   $s = 1$, $C = 0$, and b = coefficient of viscosity, or viscosity

(b) For a Bingham plastic:

   $s = 1$ and $C \neq 0$

(c) For a pseudoplastic material:

   $0 < s < 1$,   and $C = 0$

(d) For dilatant fluids:

   $1 < s < \infty$,   and $C = 0$

3.2.2. *Viscosity vs Mouthfeel Characteristics of Foods*

Most food products are non-Newtonian in fluid behavior. Hamm and Kulmbach (1967), for example, reported that the viscosity of beef muscle homogenates (finely ground *longissimus dorsi* and water mixtures) was shear-rate dependent. The shear stress-shear rate behavior was similar to that of a pseudoplastic material (Figure 4). Without citing any sensory data, Hamm and Kulmbach (1967) suggest that the proportionality factor, η, may be designated as 'dynamic toughness,' or 'viscosity.'

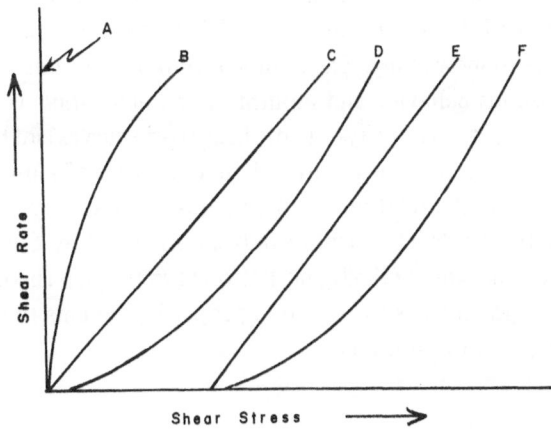

A – Ideal fluid (Schlichting, 1968)
B – Non-Newtonian, dilatant flow (Jastrzebski, 1959)
C – Newtonian, viscous flow (Alfrey, 1957)
D – Non-Newtonian, pseudo plastic flow (Jastrzebski, 1959)
E – Bingham body, idealized plastic flow (Alfrey, 1957)
F – Quasi plastic flow (Alfrey, 1957)

Fig. 4.   Rate of shear strain as a function of stress for different types of fluids.

Numerous investigations, however, have been made to relate viscosity to texture or the mouthfeel characteristics of foods. Szczesniak *et al.* (1963a), for example, developed a rating scale which related objective and sensory measurements of viscosity. Objective measurements were made with a HAT Brookfield viscometer at 100 rpm and 75 °F. Viscosity was perceived organoleptically as the force required to draw material from a spoon over the tongue. An eight-point viscosity scale, with water as a standard reference material for '1' and condensed milk representing '8', was used for the sensory ratings. The logarithm of Brookfield viscosity exhibited a slightly curvilinear relationship when plotted against equal increments of 'sensory viscosity.'

Szczesniak (1968) cited an excellent correlation ($r = 0.98$) between viscosity and sensory scores for the 'consistency' of cream soups and sauces. In this instance, the logarithm of viscosity was again found to be significantly related to equal increments of change in the mouthfeel characteristics (consistency) of the food product. Even though the product was pseudoplastic, a viscosity measurement at a shear rate of 50 s^{-1} was found to be a meaningful assessment of consistency.

Szczesniak and Farkas (1962) correlated the mouthfeel characteristics of gum solutions with their viscosity behavior. Viscosity was measured with a Brookfield viscometer at spindle speeds from 0.5 to 100 rpm. Gum solutions were rated by a trained texture-profiling panel and rated on a 7-point scale for 'degree of sliminess.' A slimy material was defined as 'one that is thick, coats the mouth, and is difficult to swallow.' Viscosity-rpm curves showed different degrees of deviation from the characteristic Newtonian type of behavior. Based upon these curves, the gum solutions were divided into three groups. The first group showed a very sharp drop in viscosity with increasing shear rate and were typically non-slimy in their mouthfeel characteristics. The third group, on the other hand, was comprised of gums that showed only a small to moderate change in viscosity with increasing shear rate. These were generally rated organoleptically as slimy to extremely slimy. The middle group of solutions was generally intermediate in both viscous behavior and mouthfeel characteristics. Hence, the study revealed that the shape of the viscosity-rate of shear (rpm) curves can be used to estimate degree of sliminess, an important mouthfeel characteristic of gum solutions.

Additional reports on the relationship between body, viscosity, and texture of gum solutions in food product development have been presented by Charm and McComis (1965), Farkas and Glicksman (1967) and Patton (1969). Such rheological studies frequently provide a rational basis for selecting proper ingredients to impart the desired textural attributes to a food product.

3.3. PLASTICITY

Some solids exhibit elastic (recoverable) deformation up to a strain where 'yielding' or non-recoverable deformation (flow) occurs under a constant level of stress. This behavior differs from viscous flow in that permanent deformation is exhibited only after stress exceeds a finite yield value. This non-elastic dissipative type of behavior is referred to as 'plastic deformation.' Theory of plasticity (Hill, 1960) deals primarily with the mathematical description of stress and strain in solids which are undergoing plastic deformation, especially during such engineering processes as extrusion of rods and tubes, drawing of wire, and rolling of strip or sheet metals.

3.3.1. *Ideal Plastic Flow*

Reiner (1960) has discussed the use of a simple model to represent ideal plastic deformation. The model (Figure 5) consists of a weight resting on a horizontal surface. To slide the weight horizontally, a force must be applied to overcome solid (dry) friction. Forces smaller than the frictional resistance have no effect. A sufficiently large critical force will cause the body to slide. When the force is released, or decreased below frictional resistance, motion will cease.

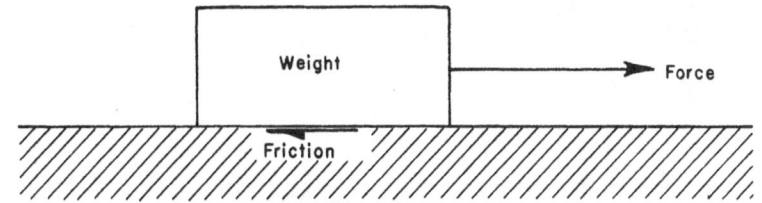

Fig. 5.   An ideal plastic model, the St. Venant body.

This model (Saint-Venant body) actually does not truly describe ideal plastic flow. Since static friction is normally greater than kinetic friction, the force required to move the weight initially will be greater than that required to maintain motion. Hence, the model simulates a material having an initial yield stress greater than that necessary to sustain plastic flow (Figure 6). This difference is usually not recognized.

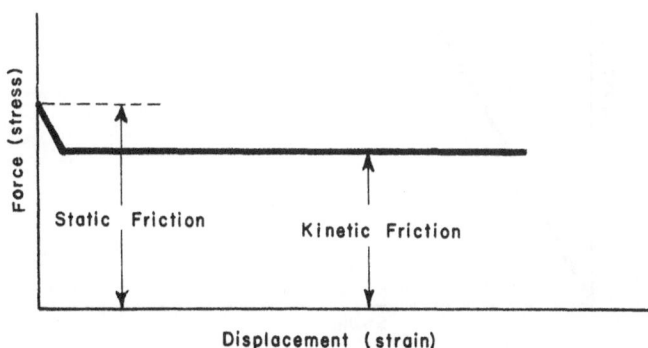

Fig. 6.   Force-displacement diagram for the St. Venant body.

Another consideration is that real materials tend to exhibit elastic as well as plastic deformation. To simulate their response, an elastic element may be added in series with the friction element (Figure 7). Finally most materials deviate from this combination of ideal elastic and ideal plastic behavior. The stress-strain relationship thus will not be linear but shows a curvilinear relationship, especially in the transition region between elastic and plastic deformation (Figure 8). Thus, the yield stress and the transition to plastic behavior may not be well defined This is true for most food products.

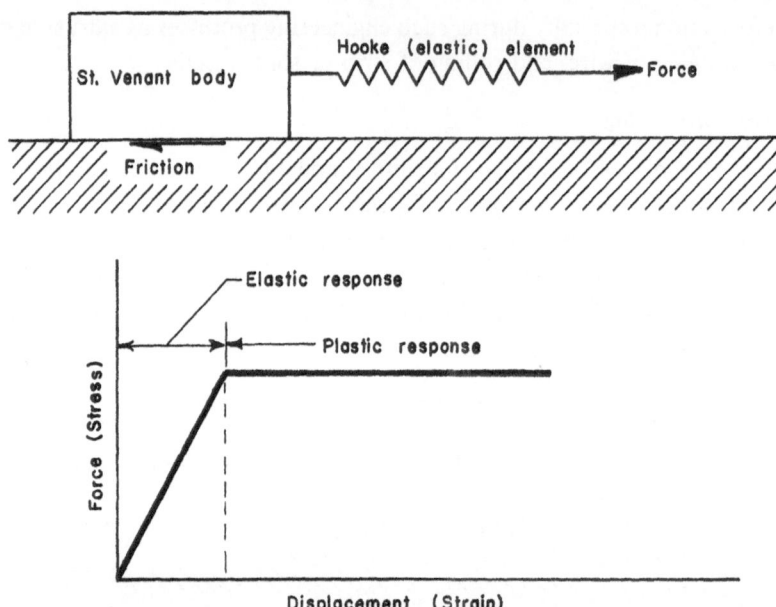

Fig. 7.   A model (top) and an idealized response diagram (bottom) for plastic materials.

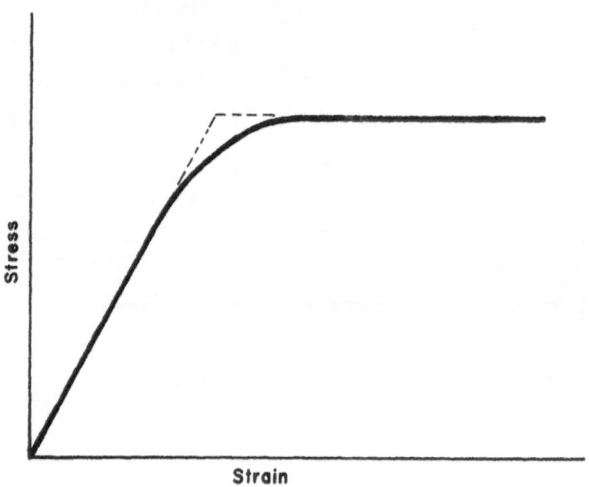

Fig. 8.   Stress-strain diagram illustrating deviation from the idealized elastic-plastic behavior.

3.3.2. *Plastic Flow of Foodstuffs*

Application of the theory of plasticity to food materials is extremely limited. Scott Blair (1953) reports examples of plastic flow in foodstuffs such as bakery doughs and dairy products. Haighton (1959) classified margarine, fats, shortenings, and butter as elasto-plastic or 'quasi-plastic' and used a cone penetrometer to measure yield stress as an index of product hardness. Finney and Hall (1967) briefly noted a similarity between the stress-strain relationship for potato tissue and the behavior of a linear strain-hardening incompressible plastic material having a negligible elastic range.

The load-displacement relationship in agar jelly (Jones, 1968) is remarkably similar to the elastic-plastic model behavior illustrated in Figure 7. The load-displacement relationship was reported by Jones (1968) to be linear up to a force of approximately 125 g. Up to this force, the ratio of force to deformation was proportional to the modulus of elasticity or rigidity of the jelly. Beyond this loading point, deformation of the jelly continued to increase even though the load (force) remained approximately constant.

Extrusion tests also have been used to assess textural characteristics of foods. Bourne and Moyer (1968), for example, applied extrusion principles to measure texture of fresh peas. Badgley (1967) patented an apparatus for comparative testing of tenderness of meat based upon its tendency to extrude during the application of pressure. Kramer and Hawbecker (1966) used an extrusion cell to measure rheological properties of gels. Bourne and Moyer (1968) cite various other tests for measuring texture that use extrusion principles.

In general, however, applications of the theory of plasticity to studies of foodstuffs have been of a cursory empirical nature. In most instances, the plastic flow of foods has been disregarded and the mechanical response has been explained from a viscoelastic model viewpoint, i.e., through a combination of the response of elastic and viscous elements.

3.4. VISCOELASTICITY

In previous sections, we have considered how concepts from elasticity and viscosity theory may be applied to characterize the mechanical behavior of certain foods. In many instances, however, foodstuffs possess rheological properties associated with both the elastic solid and the viscous fluid. In such cases, these foodstuffs may be referred to rheologically as 'viscoelastic' materials.

3.4.1. Characterizing Viscoelastic Materials

Techniques for characterizing the viscoelastic behavior of materials have been discussed by numerous authors, for example, Gross (1953), Eirich (1956), Alfrey (1957), Scott Blair and Reiner (1957), Reiner (1960), Bland (1960), and Ferry (1961). They have shown that theories developed to represent the behavior of elastic and viscous materials may also be applied to viscoelastic bodies. For viscoelastic materials, elastic moduli and other coefficients are then not considered as constants, but as material functions of time or frequency. Hence, through the use of appropriate mathematical techniques, an elastic solution may be used to develop an expression to represent the response of a viscoelastic material to stress or strain. These mathematical techniques are discussed by Gross (1953), Alfrey (1957), and Ferry (1961).

Alfrey (1957) summarized seven distinct methods of specifying properties of viscoelastic materials and divided them into two classes. In Class I, he lists theoretical approaches, i.e.:

(a) the generalized Voight (Kelvin) model;

(b) the generalized Maxwell model;

(c) the linear differential operator technique, and

(d) the complex variable mathematical approach using the mechanical impedance function.

Class II methods include the experimental curves used to 'map out' the viscoelastic character of the material, i.e.

(a) the creep curve, showing strain as a function of time at constant stress;

(b) the relaxation curve, showing stress as a function of time at constant strain; and

(c) the dynamic modulus curve, consisting of the elastic modulus as a function of the frequency of sinusoidal strain.

For linear viscoelastic materials, the three types of experimental curves should yield consistent results. That is, the moduli and coefficients from the relaxation, creep, and dynamic tests should be interconvertible mathematically. Additionally the moduli or compliances (reciprocals of the moduli) should be independent of the magnitude of the imposed stress or strain.

One technique used to describe the viscoelastic behavior of materials is in terms of mechanical models. Such models are composed of at least one of both primary elements – the elastic element and viscous element. These are sometimes referred to as the 'Hookean' (spring) element and the 'Newtonian' (dashpot) element, respectively. If for example, F is the force applied to an elastic element (Figure 9a) and u is the corresponding displacement, then

$$F = Eu,$$

where E is the spring constant (analogous to Young's modulus of elasticity in stress-strain terminology). For the viscous element in Figure 9b, the expression takes the form

$$F = \eta \dot{u},$$

where $\dot{u}$ indicates differentiation of u with respect to time, and η represents the viscosity of the dashpot.

The combination in series of these two elements (Figure 9c) forms a Maxwell model and it can be shown that the relationship between two applied force and the displacement can be represented by the differential equation

$$\dot{u} = (1/E)\,\dot{F} + (1/\eta)\,F.$$

If, on the other hand, the two primary elements are connected in parallel as shown in Figure 9d, then the resulting force-deformation relationship takes the form

$$F = Eu + \eta \dot{u},$$

which describes the behavior of a Voight (Kelvin) model. These and other more generalized models have been discussed in detail in the books cited previously, for example, Bland (1960) and Ferry (1961).

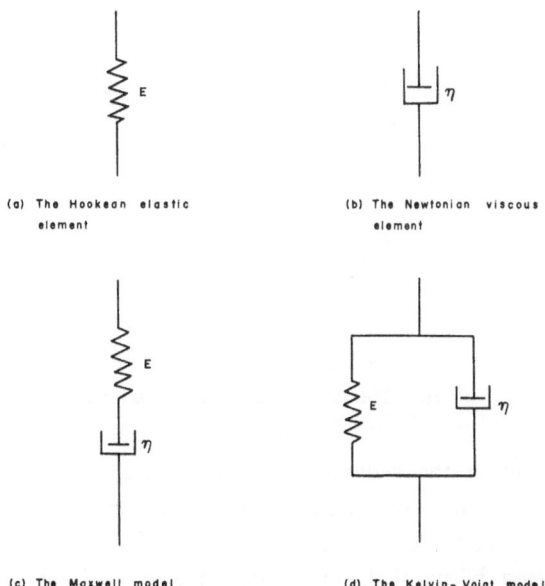

(a) The Hookean elastic element

(b) The Newtonian viscous element

(c) The Maxwell model

(d) The Kelvin-Voigt model

Fig. 9.   Some fundamental elements and mechanical models in viscoelasticity.

3.4.2. *Viscoelastic Characteristics of Foods*

The mechanical model approach has been used to describe the viscoelastic behavior of a wide range of food products, including wheat flour doughs (Lerchenthal and Muller, 1967), ice cream (Shama and Sherman, 1966), meat (Sato and Nakayama, 1970), fruits and vegetables (Mohsenin and Morrow, 1968; Finney *et al.*, 1964) as well as pea beans and grains (Zoerb and Hall, 1960, Mohsenin, 1970b). To facilitate our discussion of the results of some of these reports, consider a model consisting of a number of Maxwell models and a Hookean element all connected in parallel (Figure 10). If the model is given a deformation, u, such that $u=k$ at time $t=0$, then for $t>0$, the relaxation curve can be represented in terms of the equation

$$F(t) = kE_1 + k \sum_{i=2}^{n} E_i \exp(-E_i t/\eta_i).$$

The force response to a unit extension, excluding the constant component, is called the 'relaxation function,' which will be denoted herein as $R(t)$, i.e.

$$R(t) = \sum_{i=2}^{n} E_i \exp\left(-E_i \frac{\tau}{\eta_i}\right).$$

If the relaxation times, τ_i, are defined as

$$\tau_i = (\eta_i/E_i)$$

then

$$R(t) = \sum_{i=2}^{n} E_i \exp(-t/\tau_i).$$

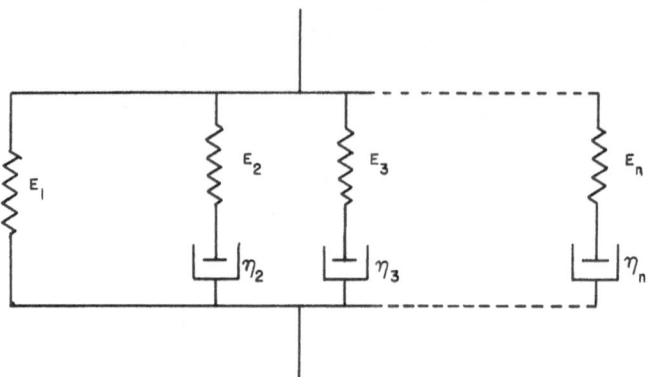

Fig. 10. The generalized Maxwell model.

This modified form of the generalized Maxwell model was used by Sato and Nakayama (1970) to describe stress relaxation within minced meat from White leghorn cockerels. Their model consisted of one Hookean (elastic) element and three Maxwell models, all connected in parallel. They suggested that the equilibrium modulus (modulus of the elastic element) and the distribution of relaxation times may be related to the binding quality of minced meat. Equilibrium modulus values were on the order of 10^4–10^5 dyne cm^{-2} and offered an indication of the presence of crosslinking within the meat after heating. Relaxation times for heated meat were shorter than those for raw meat samples and the distribution function of relaxation times varied with the type of meat evaluated. Sato and Nakayama (1970) concluded that the viscoelastic characteristics of minced meat were related to binding quality as well as to chemical composition.

Shama and Sherman (1966) studied the creep behavior of ice cream under shearing stress. For stresses up to 4000 dynes cm^{-2}, strain was linearly related to stress. Creep curves were recorded for mixes having various levels of fat content, percent overrun, and at various temperatures from -7 to $-15\,°C$. The observed creep behavior was characterized in terms of a six-element viscoelastic model consisting of a Maxwell model and two Voight-Kelvin models all connected in series. By examining the relative effect of fat, overrun, and temperature on the viscoelastic parameters associated with this model, the authors were able to deduce a relationship between structural composition and rheological behavior. Ice crystals, for example, were said to be primarily associated with instantaneous elasticity. Fat crystals, on the other hand, exerted a major influence upon the viscous elements of the model. The weak stabilizer-gel network and protein-enveloped air cells were reported to have both elastic and viscous components and tended to contribute to a behavior which is referred to as 'retarded elasticity.' Whitehead and Sherman (1967) made a subsequent study to determine if there was any significant difference in the magnitude of the rheological parameters associated with 'good' and 'poor' texture ice cream. They reported that ice cream having 'good' textural quality exhibited significantly higher values of all rheological parameters than ice cream having 'poor' texture.

In addition to the previously mentioned work on meat and ice cream, other studies have been made on the viscoelastic behavior of wheat flour doughs (Lerchenthal and Muller, 1967), fruits (Mohsenin and Morrow, 1968), and grains and seeds (Zoerb and Hall, 1960). In addition to the creep and stress relaxation approaches, a number of investigators have also used the dynamic modulus experimental technique on wheat flour doughs (Hibberd and Wallace, 1966a; Smith *et al.*, 1970), on cereal grain (Wen and Mohsenin, 1970), and on fruits (Finney, 1967; Hamann, 1969). These studies show that it is possible to apply many principles and techniques from physics to measure and characterize the viscoelastic behavior of foodstuffs. The foremost challenge however is to deduce the relationship between these viscoelastic parameters and the textural quality of foods.

4. Concluding Remarks

The difficulties associated with measuring and characterizing the rheological and textural properties of foods stem primarily from their complex and variable structure. Most foodstuffs have properties which vary from one point to another within their mass. Even at a given point within the material, the structure and properties may also vary with direction. Further, foods usually are neither perfectly elastic, plastic, nor viscous, but more generally exhibit a rheological behavior which combines these classic effects.

In spite of these complicating factors, many foodstuffs do behave in a predictable manner and concepts from the theories of elasticity, plasticity, and viscosity can be applied to obtain useful information about their textural properties. Research investigations, for example, have shown a relationship between modulus of elasticity and firmness of certain food products. Strength characteristics under shearing stress have been related to tenderness and toughness. The viscosity profiles of some products have been correlated with mouthfeel characteristics. The plastic properties of fats and the viscoelastic parameters of ice cream have been associated with textural quality. A wide variety of rheological theories and techniques, therefore, have been used to assess the textural properties of food products and it is reasonable to assume that concepts from rheology will become even more important in the future as scientists strive to better understand and characterize the texture of foods.

CHAPTER V

STRUCTURE AND TEXTURAL PROPERTIES OF FOODS

PHILIP SHERMAN

1. Introduction

Since the textural properties of foods are greatly influenced by their internal structure, a detailed study of structure and its interrelationship with textural properties should indicate how the latter can be modified and, furthermore, how to synthesize new textural concepts. With fabricated foods the possibilities would appear to be infinite (Corey, 1970b) since a wide variety of structural modification should be possible. In the case of natural foods however, e.g., meat, fruit, and vegetables, little can be done until we know more about the natural factors which affect their microstructure. Recent studies on peas (Angel and Kramer, 1969) and strawberries (Szczesniak and Smith, 1969) give some indication as to the complexity of these problems.

2. Importance of Correct Test Conditions for Evaluation of Textural Properties

The procedures for sensory evaluation of textural properties by specialized and general consumer panels are resonably well established (See Chapter III, this volume, p. 17.)

Instrumental evaluation of textural properties is a more complex problem. Two questions need to be considered before undertaking such studies. First, do we want to characterize the food for quality control purposes, and second, do we want information about the consumer's perception of textural properties. It cannot be overemphasized that the mechanical forces relevant to these two test situations may be very different. For the first situation the test conditions should be so arranged that the food undergoes minimal structural change during the test. This is achieved by applying a very small mechanical force. The test conditions required to meet the second situation are enshrouded in mystery, and all that can be said at present with any degree of certainty is that the mechanical forces applied to a food by the consumer vary with usage; e.g. cutting, stirring, shaking in a container, spreading on bread or crackers, chewing etc. A good deal of study needs to be devoted to this most important problem if foods are to be evaluated in a meaningful way in the second test situation. Far too many instrumental evaluations of textural properties are made at present utilizing arbitrarily selected force conditions, thus rendering the accumulated data of questionable value.

The importance of applying the correct forces in the second test situation can be illustrated in an elementary way by reference to fluid foods. Their texture spectrum basically involves only the property of viscosity, although stickiness or sliminess may

A. Kramer and A. S. Szczesniak (eds.), Texture Measurements of Foods, 52–70. All Rights Reserved.
Copyright © 1973 by D. Reidel Publishing Company, Dordrecht-Holland.

also have to be considered in some cases (Szczesniak and Farkas, 1962). When comparing the viscosities of two Newtonian fluids (fluids 1 and 2 in Figure 1) the shear rates under which the viscosities are measured are not important. The viscosities of the two fluids remain constant irrespective of the rate of shear at which they are measured, and fluid 1 will always be more viscous than fluid 2; however, the majority of fluid foods are non-Newtonian and their viscosities usually decrease as the rate of shear increases. There are, of course, a few exceptions to this general statement, and in such cases the viscosities increase as the rate of shear increases, but this does not affect the reasoning developed in this discussion. Suppose that we now wish to compare the viscosities of two non-Newtonian fluids (fluid 3 and 4 in Figure 2) whose viscosity-rate of shear curves intersect at a rate of shear equivalent to $X \, s^{-1}$. If we measure the viscosities of fluids 3 and 4 at a rate of shear lower than X then fluid 4 will appear to be the more viscous. On the other hand, if we measure the viscosities at a rate of shear higher than X the order is reversed and fluid 3 appears the more viscous. Further-

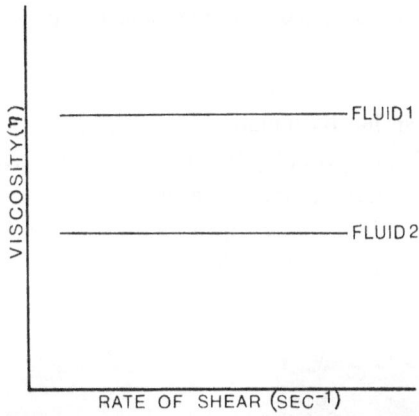

Fig. 1.   Rate of shear dependence of the viscosity of Newtonian fluids.

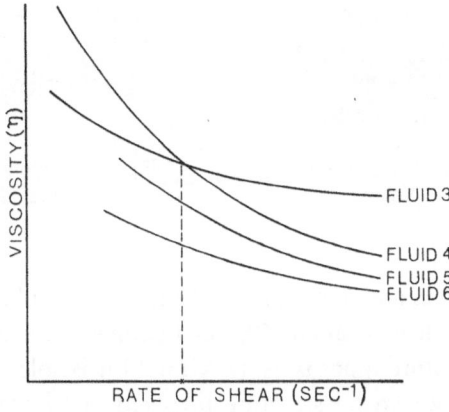

Fig. 2.   Rate of shear dependence of the viscosity of non-Newtonian fluids.

more, if we measure the viscosities at a rate of shear equivalent to X the two fluids will appear to have the same viscosity. What then is the correct rate of shear at which to measure the two viscosities? The answer to this question depends on the shear conditions to which these fluids will be subjected during use. The situation is much more complicated for solid foods because many more textural parameters are involved and this problem arises for each and every one of these parameters.

It should be mentioned that we may be fortunate in having a test situation as depicted by fluids 5 and 6 in Figure 2. Both fluids are non-Newtonian but their viscosity-rate of shear curves never cross. In this case fluid 5 will always be rated as the more viscous, but with increasing rate of shear the difference between the two viscosities decreases sharply. If the difference between the two viscosities is evaluated by members of a panel using scoring procedures then the degree of agreement with instrumental evaluation of the viscosity difference will depend on the rate of shear used in the latter case. The variability in the degree of agreement will not be important when one is concerned only with establishing which fluid is the more viscous.

3. Interrelationship of Structure and Textural Properties

3.1. ICE CREAM

The most detailed analysis has been made with frozen ice cream (Shama and Sherman, 1966; Sherman, 1966). This is probably because it has a relatively simple structure, and also because it is a manufactured and not a natural food so that its structure can be altered without too much difficulty by adjusting the formulation and/or the processing conditions. Basically, ice cream is made by freezing an emulsion of fat-in-water,

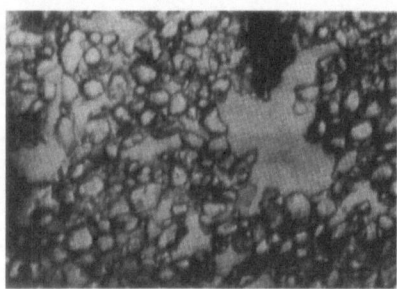

Fig. 3. Photomicrograph of ice crystal structure in frozen ice cream.

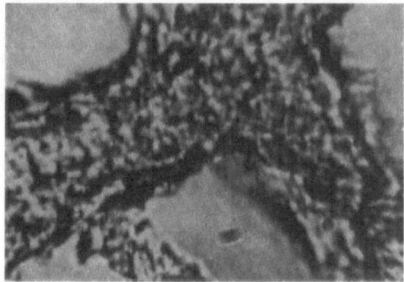

Fig. 4. Photomicrograph of fat crystal structure in frozen ice cream.

stabilized by milk protein and non-ionic emulsifier, and simultaneously incorporating a volume of air equivalent to about 50% of the final volume. The product is then stored at a low temperature (approx. $-11\ °C$) until it is sold. Figures 3 and 4 show photomicrographs of ice cream structure prepared in accordance with Arbuckle's (1960) procedure to highlight the ice crystal and fat structures respectively.

When frozen ice cream is subjected to a constant shear stress at $-11\,^{\circ}C$ in a parallel plate viscoelastometer the shear strain develops with time as shown in Figure 5. The ordinate is plotted as the ratio shear strain/shear stress or creep compliance.

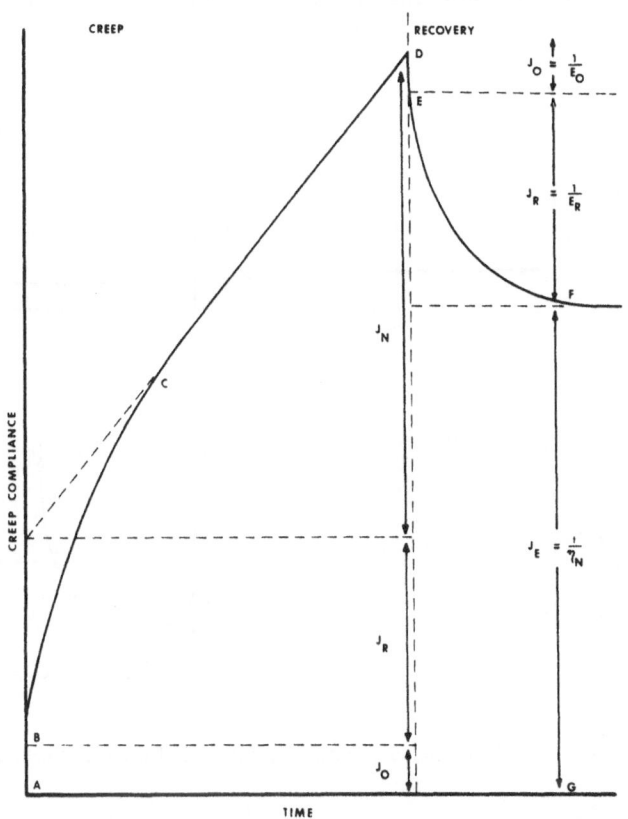

Fig. 5. Creep compliance-time curve for 10% fat ice cream at $-11\,^{\circ}C$.

Figure 5 indicates that frozen ice cream is viscoelastic (see Appendix 1) since it exhibits both elasticity (solid behavior) and viscosity (fluid behavior). The viscoelasticity is linear in form because all creep compliance-time curves superimpose on one another except at extremely large shear stresses. As shown in the Appendix, the creep compliance $J(t)$ with time depicted in Figure 5 can be represented by

$$J(t) = J_0 + \sum_i J_i [1 - \exp(-t/J_i\eta_i)] + t/\eta_N \tag{1}$$

and since compliances are the reciprocals of elastic moduli, the textural properties of the sample can be defined in terms of the physical parameters $E_0, \sum E_i, \sum \eta_i$, and η_N.

Many other foods (e.g. margarine, baked cakes, vegetables, and fruit) exhibit viscoelastic behavior when subjected to a small shear stress. This can be achieved in a parallel plate viscoelastometer, in compression, or in extension.

In order to establish the interrelationship between the parameters quoted in Equation (1) and ice cream structure, the latter was altered by varying, in turn, the fat content between 0 and 10% at -11 °C, the air volume between 1% and 55% for a 10% fat ice cream, and the test temperature between $-11°$ and $-15°$C for a 10% fat ice cream. The data showing how the physical parameters were affected are given in

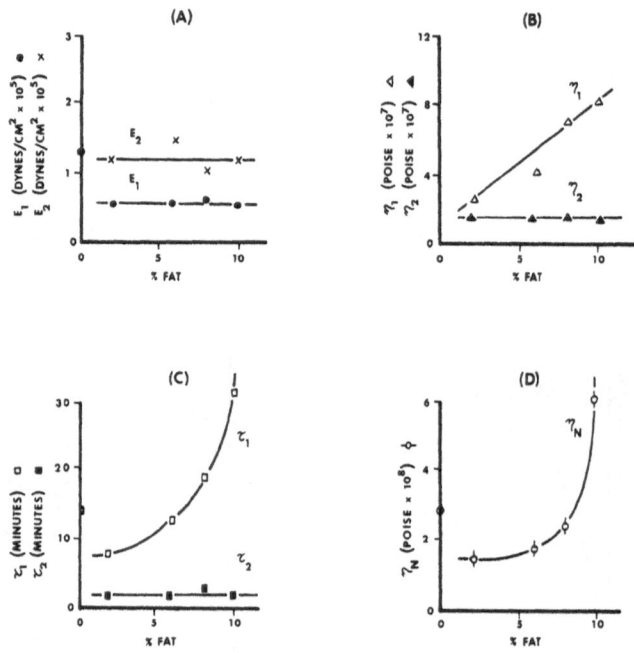

Fig. 6.   Influence of % fat on the viscoelastic parameters of frozen ice cream at -11°C.

Figures 6, 7, and 8. It is interesting to note that the values of all parameters are several orders of magnitude lower than for pure ice (Jellinek and Brill, 1956). It was found in all tests that $\sum E_i$ and $\sum \eta_i$ could be represented satisfactorily by E_1, E_2, and η_1, η_2 respectively.

Table I summarizes the main findings of the data contained in Figures 6, 7, and 8. These data, along with our knowledge about the microscopic structure, suggest that frozen ice cream can be regarded as an aerated ice crystal structure which has been modified in textural characteristics by the superimposition of a fat crystal network. Milk protein films of polymolecular thickness envelop the air cells and fat crystals, while the supercooled fluid matrix contains sugar, electrolytes, hydrocolloid additives etc. Although the ice crystals are relatively large (up to approx. 50μ) there are few points of contact between them (Figure 3), and the separation distances are usually a few microns; therefore, there is little possibility of linkages being established between the ice crystals. On the other hand, there is strong microscopic evidence (Figure 4) for aggregation of fat crystals (0.5–4.0 μ diameter) both around the air cells and in the liquid matrix, so that these crystals could provide the locus of linkage formation.

Referring now to Table I, we find that the parameters E_1 and E_2 are influenced by air volume and temperature but not by fat content. At a test temperature of $-11\,^\circ$C about 75% of the water in ice cream is frozen, and this percentage increases as the temperature is lowered still further. This means that the amount of unfrozen water which can associate with the hydrocolloid additive to form a gel decreases with decreasing temperature, so that a more concentrated, and consequently a stiffer, gel is

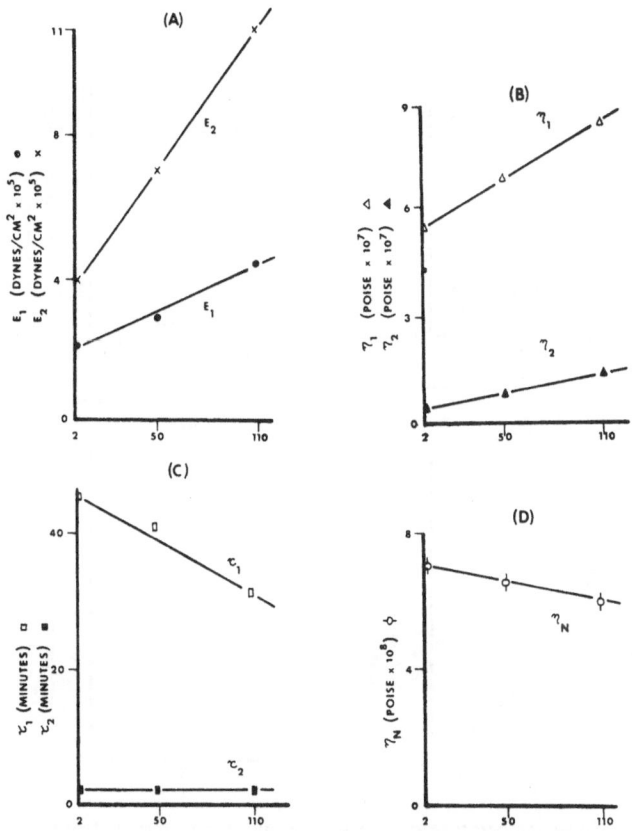

Fig. 7.   Influence of air volume on the viscoelastic parameters of frozen 10% fat ice cream at $-11\,^\circ$C.

progressively formed within the matrix. When the air volume is increased, the air cells, ice crystals, and fat crystals become smaller (Shama and Sherman, 1966), i.e. their number per unit volume of ice cream increases. More linkages are established by the fat crystals and a stronger, more interlinked, structure is developed. This is reflected in the rise in E_1 and E_2, and especially in the value of the latter parameter. The parameters η_1 and η_2 are also affected by air volume to some extent, but temperature has a much greater effect, especially on η_1. Fat content exerts a large influence on η_1, but it has no effect on η_2. It appears, therefore, that η_1 is mainly related to the interlinked fat crystal network and hydrocolloid gel structure, whereas η_2 is related to the hydrocolloid gel and air cell structures. Fat content is the only variable that exerts a large effect on η_N. The main effect appears when the fat level is about 5%, and it is probably due to increased interaction between non-linked fat crystals during viscous flow.

The interdependence of the physical parameters and the structural elements of frozen ice cream can be depicted very conveniently utilizing a mechanical model. In this model the instantaneous elasticity, retarded elasticity, and Newtonian viscosity

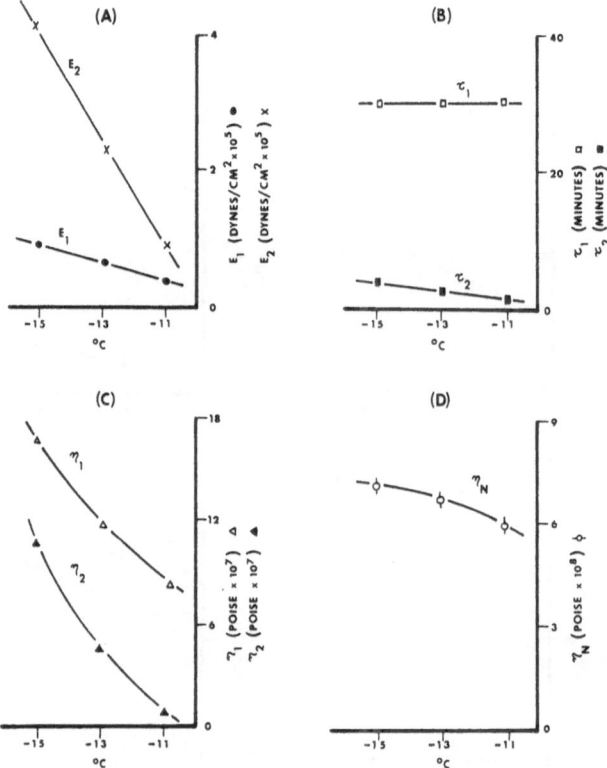

Fig. 8.   Influence of test temperature on the viscoelastic parameters of frozen 10% fat ice cream.

are represented by an elastic spring, a series of springs and dashpots in parallel (Kelvin-Voigt elements), and a dashpot respectively. Figure 9 represents a model which satisfactorily depicts the shear behavior with time, based upon the data of Figures 6, 7, and 8. It also indicates the components of the frozen ice cream structure which mainly influence the various parameters.

The mechanical model can be utilized to predict what physical parameters are important to the consumer's evaluation of textural properties. The reasoning is as follows. When ice cream melts in the mouth ice crystals melt, and air held originally within the frozen structure is released. The hydrocolloid gel structure is weakened due

TABLE I

General summary of data contained in Figures 6–8

Test condition	Instantaneous elastic compliance region	Retarded elastic compliance region				Newtonian region
	E_0	E_1	E_2	η_1	η_2	η_N
0–10% fat	decreases with increasing % fat	not influenced by % fat	not influenced by % fat	increases with increasing % fat	not influenced by % fat	increases with % fat particularly above 5% fat
1–55% air (by volume)	decreases with increasing air volume	increases slightly with increasing air volume	increases with increasing air volume	increases with increasing air volume	increases slightly with increasing air volume	decreases with increasing air volume
test temperature of $-11°$ to $-15°$C	increases with decreasing temperature	increases slightly with decreasing temperature	increases with decreasing temperature	increases with decreasing temperature	increases with decreasing temperature	increases slightly with decreasing temperatures

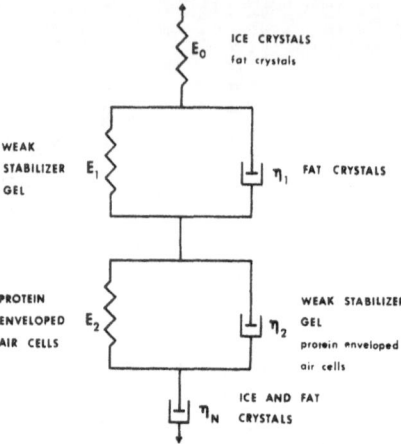

Fig. 9.   Six element mechanical model for frozen ice cream.

to dilution with the additional water which now becomes available. Fat crystals also melt, and coagulation of fat particles, which began in the ice cream emulsion in the initial stages of freezing following rupture of some of the protein films around fat particles (Sherman, 1965), is no longer prevented. This weakens the interlinked fat particle network which was present in the frozen structure. Forces of attraction (probably London-van de Waals' forces) between the residual liquid fat particles will be weaker than those existing between solid fat crystals in the frozen ice cream. These various changes indicate that E_0, E_1, η_1, and η_N in Figure 4 should all decrease substantially, and that E_2 and η_2 could well disappear. Consequently, it should be possible to represent the strain behavior of thawed ice cream at a constant low shear stress by a mechanical model with only 4 elements (Figure 10). This view has been confirmed by studies with a coaxial cylinder viscometer (Sherman, 1966).

Following the creep compliance-time studies on thawed ice cream, referred to in the previous section, similar studies were made at 21 °C on a range of ice creams rated as having 'good' or 'poor' texture by a consumer panel (Whitehead and Sherman, 1967). No attempt was made in these tests to apply texture profiling methodology. The purpose of these tests was to establish whether all the physical parameters shown in

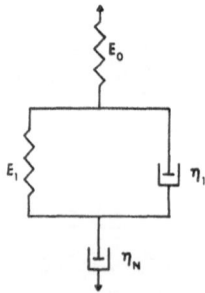

Fig. 10.   Four element mechanical model for thawed ice cream.

Figure 10 contribute to the concept of 'good' texture or whether certain parameters are more important than others. Typical data are shown in Table II. It is readily seen that the values of all four physical parameters are significantly higher for the 'good' texture samples, with the differences being particularly marked for η_1 and η_N.

TABLE II

Rheological parameters of 'good' and 'poor' texture ice creams after melting (Whitehead and Sherman, 1967)

Texture quality	Sample No.	E_0 (dyne cm^{-2})	E_1 (dyne cm^{-2})	η_1 (poise)	η_N (poise)
'good'	1	143	156	5.2×10^3	3.7×10^4
	2	416	278	6.6×10^3	5.6×10^4
	3	793	299	13.3×10^3	8.6×10^4
'poor'	4	62	33	2.6×10^3	2.3×10^4
	5	21	14	0.7×10^3	1.1×10^4
	6	42	25	1.9×10^3	1.0×10^4

On the basis of the discussion in the Introduction it could be argued that these thawed ice creams should have been subjected to much greater shearing forces during these tests to simulate mechanical action in the mouth. The rates of shear applied were 10^{-3} to 10^{-4} s^{-1}, whereas the magnitude is undoubtedly much higher in the mouth (Wood, 1968; Shama and Sherman, 1970). Presumably the choice of samples was fortunate in that the respective curves for any one physical parameter always fell in the same order and they never intersected, i.e. we have a situation similar to that for fluids 4, 5, and 6 in Figure 2.

3.2. MARGARINE AND BUTTER

Less detailed studies of the above type have been made on margarine and baked cakes. The primary aim with these materials was to follow the changes in textural properties during storage following manufacture, and to deduce what structural components could be responsible for these changes. Analysis of creep compliance-time curves for margarine, which was stored and tested at 15 °C over a 35-day period, indicated that the behavior could be represented by Equation (1) in the form

$$J(t) = J_0 + J_1 \left[1 - \exp\left(-t/J_1\eta_1\right)\right] + \\ + J_2 \left[1 - \exp\left(-t/J_2\eta_2\right)\right] + J_3 \left[1 - \exp\left(-t/J_3\eta_3\right)\right] + \\ + J_4 \left[1 - \exp\left(-t/J_4\eta_4\right)\right] + t/\eta_N \tag{2}$$

so that the textural properties could be defined by 10 physical parameters viz. E_0, E_1, E_2, E_3, E_4, η_1, η_2, η_3, η_4, and η_N (Shama and Sherman, 1968). The value of each parameter increased very rapidly during the initial stages of storage and then asymptotically to an approximately steady value. This was attributed to continued growth of fat crystals and the establishment of more strongly interlinked networks of fat crystals causing the margarine to become much harder.

Sone (1961) has observed similar trends in butter stored at 20 °C after it has been worked for 30 min on a roller mill. If continued growth of fat crystals during the storage period occurs around randomly distributed centers then it is possible to analyze the kinetics of textural change by Avrami's (1939, 1940, 1941) theory. Using the Newtonian viscosity data the kinetics of fat crystal growth can be represented by

$$\ln \ln \left(\frac{\eta_{N(\infty)} - \eta_{N(t)}}{\eta_{N(\infty)} - \eta_{N(0)}} \right) = -(n \ln t + \ln k), \tag{3}$$

where $\eta_{N(0)}$, $\eta_{N(t)}$, and $\eta_{N(\infty)}$ are the η_N values initially, at time t, and after an infinitely long time, k is a constant which depends on the nucleation mechanism and the rate of crystal growth, and n has a value between 1 and 4 depending on the nucleation mechanism. A plot of the left hand side of Equation (3) against $\ln t$ gives a straight line with a gradient of n and an intercept equal to $\ln k$. Sone (1961) found that n was approx. 1.2, which suggests that the fat crystals had a needle-like shape at 20 °C. At lower temperatures the crystals appeared to be in the form of small plates. Fat crystals extracted from unworked butter and margarine stored at 5 °C have axial ratios of approx. 3.5 (Parkinson et al., 1970).

3.3. BREAD AND BAKED CAKES

Staling of bread and of baked cakes at room temperature (21° –25° C) follows a trend similar to the post-manufacture hardening of margarine, although the time span involved is only 3–4 days. Cornford et al. (1964) analyzed the staling process using Equation (3) but with viscosity values replaced by values of the crumb modulus (stress/strain ratio derived from readings of the weight required to compress samples by 50% in 1 min). The data suggested that one of the main factors responsible for staling is the development of a more ordered arrangement of molecules, probably as a result of starch crystallization. A significant observation in the staling of baked cake (Shama and Sherman, 1968) is that in the fresh state it exhibits non-linear viscoelasticity, i.e. the strain/stress ratio increases with increasing stress. After a few days staling the behavior remains linear for shear stresses up to 8000 dyne cm^{-2}, which indicates some significant alteration within the structure. This change must be attributed to structural alterations different from those found in bread because, whereas elasticity becomes increasingly important during staling of bread and viscosity becomes progressively less important, in cake the relative contribution of the physical parameters does not alter to anything like the same degree.

3.4. FLUID FOODS

The factors which influence the rheological properties of emulsions and dispersions of solid particles in fluid media have now been reasonably well identified (Sherman, 1970). This knowledge can be utilized to build up a picture of structural factors that will influence the sensory evaluation of viscosity and consumer acceptability of products such as fruit juices and purees, salad dressings, dairy products etc. Margarine is a product to which this approach has already been applied (Parkinson *et al.*, 1970).

3.5. FRUIT AND VEGETABLES

Creep compliance-time studies on apples (Morrow and Mohsenin, 1966; Mohsenin and Morrow, 1968) suggest behavior in accordance with

$$J(t) = J_0 + J_m [1 - \exp(-t/J_m \eta_m)] \tag{4}$$

thus suggesting the absence of a Newtonian viscosity contribution. This may have been due to the short time duration of the tests so that the Newtonian compliance region was not reached. No detailed analysis of the retarded elastic contribution to the total creep compliance has been reported and only the mean compliance (J_m) was given. Values of the moduli calculated from Equation (4) were greater for the intact fruit than for peeled fruit, suggesting that the outer skin plays an important role in the 'stiffness' of the fruit. The skin appears to exert its greatest influence on η_m, the viscosity associated with the retarded elastic compliance. The 'stiffening' contribution of apple skin has been confirmed (Abbott *et al.*, 1968a, 1968b) by acoustic vibration studies. Additional studies by the latter technique on samples of apple flesh taken in succession at different depths suggested that the layer of flesh just below the skin has different 'stiffening' properties from the rest of the apple flesh. Following harvesting the 'stiffening'action of the skin decreases slowly.

No detailed relationships have been established yet between the viscoelastic properties of whole fruits and vegetables and their cellular structures but some information has been derived from elongational and cyclic compression-decompression tests. A recent review by Reeve (1970) contains much valuable information that is relevant to this problem. Cellulosic microfibrils are the framework material of the cell walls and the spaces between them are filled with a large variety of encrusting materials, which are usually amorphous, and include pectins, hemicelluloses, lignins, suberins, cutins, etc. according to the type of tissue. An analogy can be drawn between the cell wall structure and that of artificial laminates in which different layers and binding materials combine to provide strength and rigidity as well as resistance to deformation (Reeve, 1970).

Somers (1965) suggests that plant cell walls are mainly responsible for viscoelastic behavior. Both the slender cytoplasmic connections between cells and the fluid protoplasm within the cells are discounted as contributors to the viscoelasticity. Interaction of the hydrostatic pressure (turgor) of the cell contents with the cell walls does, however, influence the viscoelasticity. Cyclic compression and decompression of potato tuber results in a progressive increase in stress as the strain increases, and this leads to apparent values of Young's modulus which are approximately four times as large as the values calculated for the initial stages of the cycle. These changes are attributed to rearrangements within the cellulose microfibril networks of the cell walls. Freezing or acetone treatment of strips of plant tissues results in low apparent values of Young's modulus in short time elongation tests because the semipermeable properties of the cell walls are destroyed.

Stringiness, as in raw vegetables such as celery, arises from the thickening of cell walls by the deposition of mechanical or supporting tissues. These latter tissues are collenchyma, sclerenchyma, and xylem. Cells of collenchyma and sclerenchyma are long and tapered, but whereas sclerenchyma cell walls are almost uniform in thickness, collenchyma cell walls exhibit irregular longitudinal or fluted thickenings. The wall thickenings of collenchyma contain much pectic material. Sclerenchyma fibres appear to be elastic when stretched mechanically whereas collenchyma fibres remain permanently extended.

Pectins or starches are found in many vegetables, and the changes which they undergo at cooking or processing temperatures profoundly influence the textural properties of the vegetables. The interplay between pectins and cellulosic components of the cell wall has a great influence on the textural properties of apples (Nelmes and Preston, 1968) and on the juice from tomatoes (Whittenberger and Nutting, 1957, 1958). The softening of parenchymatous fruits during ripening results from changes effected on pectins by pectic enzymes, and also changes in the hemicelluloses and lignin of the cell wall. The starch contained in potatoes swells and gelatinizes within the cells at temperatures ranging from just below 60 °C to above 70 °C depending on the size of the starch granules. As the starch swells the cell walls distend so that the original polyhedral shapes become more spherical and the cells are then pushed apart (Reeve, 1954a,b; Sterling, 1955). Starch gelatinization is followed by some breakdown of middle lamellae pectins between the cell walls of adjacent cells. This process occurs earlier in potatoes that give a 'mealy' texture when cooked. Excessive rupturing of cell walls give rise to a large amount of extracellular starch, and this produces stickiness or gumminess in mashed potato. In addition to the changes on the starch content of potatoes induced by cooking the hydrogen bonds of pectic and other cell wall polysaccharides are disrupted and cell separation follows (Sterling, 1963).

The studies in depth which are required to clarify texture properties – food structure relationships are exemplified by a recent study (Szczesniak and Smith, 1969) on strawberries. The fresh fruits were evaluated instrumentally by the General Foods Texturometer (Szczesniak, 1963a; Friedman et al., 1963), sensory evaluation was by a texture profile panel (Brandt et al., 1963), and a microscopic examination was also made. Table III summarizes the data for different stages of ripening. Fresh strawberries are firm, moderately crisp, and low in cohesiveness. As ripening proceeds they become less firm, less crisp, slightly adhesive to the touch, and juice is released more readily. The decrease in firmness is associated with degradation of the epidermal (outer) cell walls. Simultaneously the cells elongate due to enlargement of the underlying cortex tissue. The latter process occurs at different rates in different cells and it leads to a thinning of the cell walls, hence, the fragility of overripe strawberries. Pith cells, in the center of the fruit, also grow larger during ripening and in the overripe berry they are usually larger than the cells of the cortex. There is some evidence for folding of cell walls and cytoplasmic disarrangement in both cortex and pith. This contributes to the

TABLE III

Textural properties of fresh strawberries (after Szczesniak and Smith, 1969)

Evaluation procedure	Parameter	Underripe fruit	Ripe fruit	Overripe fruit
instrumental – General Foods Texturometer	firmness (g)	1470	1380	1180
	cohesiveness	0.14	0.16	0.15
	crispness (g)	1550	1160	1000
	adhesiveness (g.cm)	4.1	8.0	3.2
	% showing adhesiveness	55%	100%	83%
	juiciness (cm^2)	58.0	64.5	67.1
sensory-texture profile panel	firmness		decreasing	→
	crispness		decreasing	→
	juiciness		increasing	→
	% samples showing adhesiveness		increasing	→
microscopy	cells		elongation and expansion	→
	cell contents		liquefaction	→
	cell walls		thinning and folding	→
	cytoplasm		disarrangement	→
	pectin		degradation	→
	xylem vessels		thickening and deposition of liquid	→

reduction in crispness and firmness while pectin degradation is responsible for the decrease in cohesiveness.

A similar approach was used to compare fresh, frozen, and frozen-dried rehydrated strawberries, (Szczesniak and Smith, 1969). Table IV presents the instrumental and microscopical evaluation data. The most striking differences in the data are the drastically reduced firmness and crispness shown by frozen and freeze-dried rehydrated fruits in comparison to fresh fruit. This is associated with plasmolyzed cells, disorganized and clumped cytoplasm, folding of cell walls and cell surfaces, and cell rupture, particularly in the freeze-dried rehydrated fruits. The latter fruit was far less juicy than frozen fruit, and it lost water less readily. Staining techniques revealed changes in the cellulose, pectin, and lignin structures.

A sophisticated analysis (Angel and Kramer, 1969) has been made of histological factors which influence the texture of pea testa (skin) under uni-directional and hydrostatic pressures, and in shear. The pressures employed were similar in magnitude to those used in mastication and during general commercial handling of peas. Peas have a thin, smooth, waxy cuticle on the surface of the testa. Underlying this (Figure 11) are closely packed elongated cells (60–100 μ) with thickened walls which are called macrosclereids. Below these cells is a single layer of triangularly arranged bone-shaped cells, the osteoclereids (25–40 μ).

TABLE IV

A comparison of the textural properties of three different types of strawberries (after Szczesniak and Smith, 1969)

Evaluation procedure	Parameter	Fresh (Sparkle)	Frozen in syrup (Fancy Northwest)	Freeze-dried rehydrated (Shasta)
instrumental; General Foods Texturometer	firmness (g)	1380	290	290
	cohesiveness (g)	0.16	0.44	0.35
	crispness (g)	1160	170	0
	adhesiveness (g.cm)	8.0	6.2	10.6
	juiciness (cm²)	64.5	75.0	43.0
microscopy	air bubbles in tissue	few	many	many
	cell organization	high	some disorientation	complete disorganization
	cell state	turgid	plasmolyzed	plasmolyzed
	cell wall	intact, thin	folded, some rupture	much rupture, degradation of cellulose
	cytoplasm	fairly dense	disorganized, clumped	discontinuous
	xylem	normal	normal	leaking of liquid into adjacent cells.

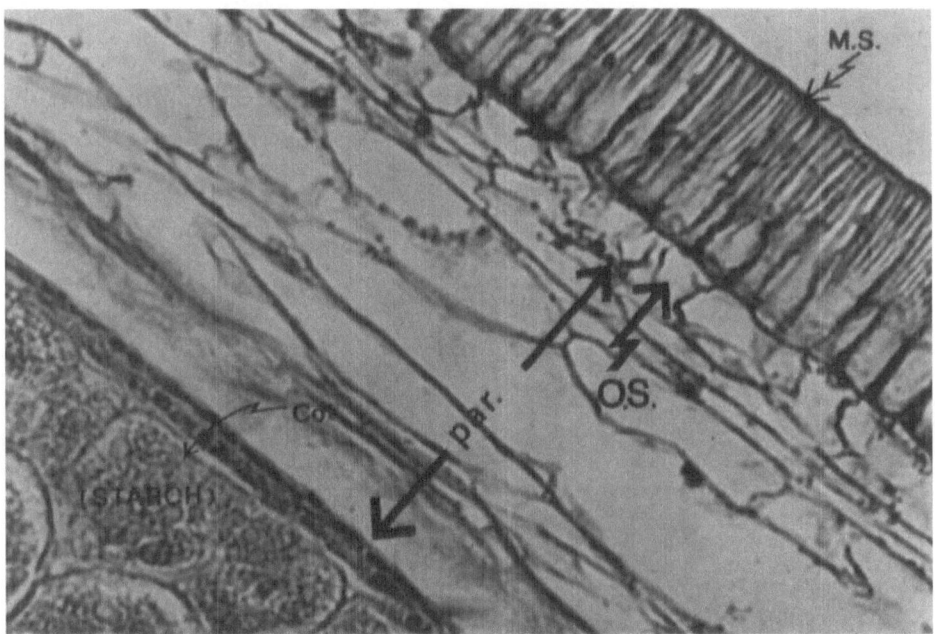

Fig. 11.   Section through pea testa, Telephone variety showing macrosclereids (M.S.), osteosclereids (O.S.), parenchyma (par.) starchy cotyledon (cot.) (Angel and Kramer, 1969).

The elasticity observed in force-deformation tests is attributed to the support offered by the osteosclereid layer to the macrosclereids which permits compressive and conconcussive forces to be absorbed by the deformed (bending) macrosclereids. Both unidirectional and hydrostatic pressures cause bending of the macrosclereids and flattening of the osteosclerieds. In addition, unidirectional pressure causes cell separation in the macrosclereid layer. Shear produces bending, buckling, and wedging of macrosclereids and this may be asociated with a pentosan – cellulose complex which is present in the basal and apical regions of these cells, and also in the wall regions of the osteosclereids. This complex gives rise to an elastic-like matrix which is responsible for the resilience and toughness of the testa.

When tomatoes ripen the mechanical characteristics of the structural components of the pericarp change in different ways. The pericarp consists of the skin (exocarp), the flesh (mesocarp), and the inner skin (endocarp). Penetration studies using an Instron Universal Tester with a plunger attachment (Holt, 1970) give force-distance traces with several force peaks (Figure 12). The main peak characterizes the passage of the plunger through the exocarp; a plateau is obtained in the mesocarp region, and this is followed by a smaller peak in the endocarp. Finally, a plateau is obtained as the plunger enters the locular juice. A greater force is required to pierce the skin of unripe than of ripe tomatoes. Both the endocarp peak and the mesocarp plateau occur at higher force values in unripe tomatoes. If the ripening process is subdivided into six phases viz. 1. green, 2. early colour break, 3. late colour break, 4. early ripe, 5. ripe, and 6. over-

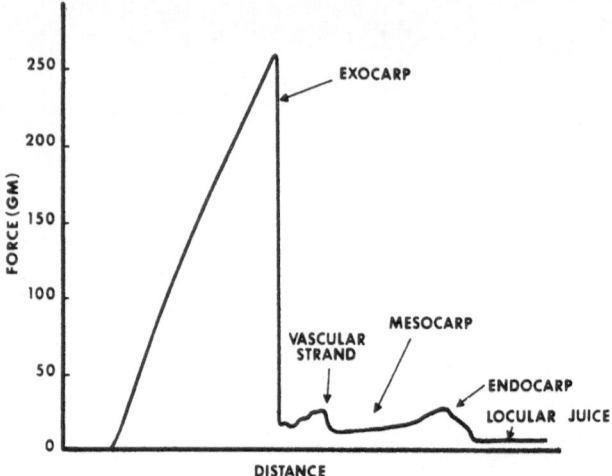

Fig. 12.   Instron force-distance curve for a ripe tomato (Holt, 1970).

ripe, the mechanical strength of the exocarp falls progressively during ripening, whereas both the mesocarp and endocarp exhibit a rapid decrease in strength during phases 1–3 and a much slower decrease in the remaining phases.

3.6. MEAT

The factors which influence the textural properties of meat have not been identified with the same degree of certainty as for some other foods (Szczesniak and Torgeson, 1965; Lawrie, 1968). Meat consists basically of muscle fibre bundles lying parallel to one another in an elastic-like matrix of connective tissue. Interactions involving the meat proteins, both prior to and after slaughter of the animal, have a large influence on the textural properties but they are not the only factors involved. For example, pre-slaughter factors such as muscle and animal characteristics, and post-slaughter factors such as changes during rigor mortis, aging, storage, and cooking are also important (Lawrie, 1968).

For convenience Lawrie (1968) subdivides meat proteins into three groups. These are the proteins of the connective tissue, of the myofibrils, and of the sarcoplasm.

The main proteins of connective tissue are collagen and elastin, but the content of elastin is relatively low. Collagen consists of a large number of rod shaped macromolecular units, each containing three helically interwoven polypeptide chains. There is an approximate inverse relationship between meat tenderness and connective tissue content, but the true situation is more complex since the degree of crosslinking of the polypeptide chains in the collagen chains also influences tenderness. So does deposition of intramuscular fat within the collagen network as it leads to a more open network structure. Szczesniak and Torgeson (1965) have summarized the published data on tenderness and connective tissue content, and they point out that the correlations are statistically more significant for those tests in which tenderness was evaluated instrumentally with the Warner-Bratzler shear press than for those tests in which it was evaluated sensorily by panels.

The myofibrils, which contain more than half the protein content of meat muscle, are longitudinally repeated structural units embodying the sarcomere. The latter is bounded on either side by the Z lines which have a central region of parallel thick filaments of the protein myosin. From the Z lines thin filaments of the protein actin extend toward the centre of the sarcomere and penetrate between the myosin filaments. When muscle contracts, and this may occur post mortem as well as 'in vivo', the actin filaments pull the Z lines inwards from each side of the sarcomere so that the degree of contraction is related to the degree of interdigitation of the actin and myosin fibres. If muscle is removed from the carcass at the beginning of rigor mortis it proceeds to shorten. Contractions to a maximum of 40% of the original length produce more cross bonding, in addition to greater interdigitation, and this results in increased toughness when the meat is cooked.

Raising the temperature, as in cooking, has opposite effects on the proteins of connective tissue and of the myofibrils. Collagen degrades, much of it being converted into gelatin at the cooking temperature, and this transition increases the tenderness of meat. On the other hand the coagulation and degradation of myofibrillar proteins reduces the tenderness. Consequently, different muscles react differently when cooked, and their ultimate textural properties depend upon the temperature-time conditions during cooking (Szczesniak and Torgeson, 1965).

Sarcoplasm, or cytoplasm, constitutes the fluid phase of muscle. It contains soluble proteins and a suspension of intercellular particles such as mitochondria and hysosomes. Denaturation of the sarcoplasmic proteins post mortem may have some effect on meat tenderness, and if so then the increase in tenderness observed during conditioning could be due largely to the sarcoplasmic proteins since they are the main proteins to break down at this time into peptides and amino acids.

Since meat is a very important commodity a multidisciplinary study of its textural properties, which incorporates methodologies such as those employed for strawberries (Szczesniak and Smith, 1969) and for peas (Angel and Kramer, 1969), could produce very pertinent information.

Appendix

The creep compliance-time curve (Figure 5) can be subdivided into five distinct regions, three of which characterize the response to the shear stress while the other two characterize the recovery following removal of the shear stress. In the region AB the sample exhibits pure elastic response, so that if the shear stress had been removed at, or before B, the sample would have fully recovered its original structure. Beyond B the rate of strain decreases because linkages between micro-units of the ice cream structure begin to rupture. The weakest linkages rupture first, and stronger linkages rupture at longer times. At C flow commences between those micro-units which are no longer linked in to the main structure. When the shear stress is removed at D, the recovery is initially instantaneous (DE), and then it decays curvilinearly (EF). The whole of the original structure is never recovered, and the distance FG is a measure of the structural damage to the sample.

The region AB represents the instantaneous elastic compliance.

Mathematically this is represented by

$$J_0 = \varepsilon_0(t)/p = 1/E_0, \tag{5}$$

where $\varepsilon_0(t)$ is the shear strain at very short times (t_0), p is the applied shear stress, and E_0 is the instantaneous elastic modulus.

The region BC represents the retarded elastic compliance J_R, where

$$
\begin{aligned}
J_R = \varepsilon_R(t)/p &= J_m[1 - \exp(-t/\tau_m)] \\
&= \sum_i J_i[1 - \exp(-t/\tau_i)] \\
&= \sum J_i[1 - \exp(-t/J_i\eta_i)]
\end{aligned}
\tag{6}
$$

and $\varepsilon_R(t)$ is the shear strain at times not exceeding t_i; $\sum J_i(=1/\sum E_i)$ is the sum of the i compliances which contribute to J_R; $\sum E_i$ are the respective elastic moduli; τ_i are their retardation times, and η_i are their associated viscosities. As mentioned in the text the linkages between the structural microunits which rupture in this region do so at different rates so that we have a spectrum of rupture times.

Finally, CD (Figure 5) is the Newtonian compliance J_N region,

$$J_N = \frac{\varepsilon_N(t)}{p} = \frac{t_N - t_i}{\eta_N}, \tag{7}$$

where $\varepsilon_N(t)$ is the shear strain in the time region $(t_N - t_i)$, and η_N is the Newtonian viscosity; therefore, we can describe the total strain of the sample with time, when a constant shear stress is applied, by the sum of Equations (5)–(7). The total creep compliance $J(t)$ is given by

$$
\begin{aligned}
J(t) &= J_0 + J_R + J_N \\
&= J_0 + \sum_i J_i[1 - \exp(-t/J_i\eta_i)] + t/\eta_N.
\end{aligned}
\tag{8}
$$

The way in which the various parameters in Equation (6) are derived has been described elsewhere (Sherman, 1970a, b).

CHAPTER VI

INSTRUMENTAL METHODS OF TEXTURE MEASUREMENTS

ALINA SURMACKA SZCZESNIAK

1. Introduction

Research on instrumental methods of measuring food texture is not a new develop-ment. In 1905 Hankoczy designed an apparatus for measuring the strength of gluten (Brabender, 1965), and in 1907 Lehmann described two instruments for testing the tenderness of meat (Lehmann, 1907). Owing to these early beginnings, the fields of dough rheology and meat texture have enjoyed the greatest amount of activity in developing instrumental test methods. In time, this interest spread to other foodstuffs such as cheeses, fruits, butter and margarine, vegetables, and – more recently – to convenience food products.

In the past, most of this work aimed at quantifying in a reproducible manner the effects of variables such as raw materials, processing, handling, and growing condi-tions. With the present realization of the importance of texture in consumer accep-tance, an increasing amount of attention is being paid to correlating instrumental with sensory methods of texture evaluation. This, coupled with general advances in instrumentation and electronics and a critical examination of the existing methodology, has brought about new trends and refinements.

A very large number of texture measuring instruments have been designed and described in the literature; several are available commercially. A list of better known named apparatuses designed by various workers is shown in Appendix 1, together with literature references where a detailed description of the design may be found. Appendix 2 lists some of the commercially available equipment and respective manufacturers or distributors.

Much attention and discussion has been devoted recently to what actually is being measured by these devices. This has been prompted by frequent poor correlations shown between instrumental and sensory evaluations and by the desire to interpret the measurement in terms of psychorheological principles. Refinements in instruments design and careful control of sample size and test conditions often contribute to improving the correlations; the most important point, however, which must be kept in mind is that texture is not 'one thing' but a spectrum of parameters. Each of the instrumental devices detects only a portion of that spectrum, some a larger portion than others. It is only the human being who can perceive, analyze, integrate, and interpret the entire spectrum of textural (and other) characteristics in one evaluation. Moreover, it is generally agreed that the concept of texture is meaningful only when viewed as an 'interaction of the human with the mechanical properties of the material' (Corey, 1970a). Thus, in the true sense, none of the instrumental devices measures texture per se, only physical properties which can be related (directly or indirectly) to textural attributes sensed organoleptically.

Sensory evaluation is faced with many methodological, psychological, and physiological problems. It is also time consuming and costly. Moreover, to many people the results still have the connotation of an 'opinion'. Instrumental measurements, on the other hand, are more apt to be regarded as 'facts' to be used rather than challenged.

Speed, reproducibility, relative ease of standardization and freedom from problems indigenous to sensory assessment have prompted the burst of activity in instrumental methods of texture evaluation. These have been reviewed by Finney (1969a), Gordon (1967, 1969), Heiss and Witzel (1969), Szczesniak (1963a, 1966), and others. Voisey (1971) has summarized the application to texture testing devices of recent developments in the field of instrumentation. A detailed review of instruments used for testing meat was published by Szczesniak and Torgeson (1965), while Kramer and Twigg (1970) treated the subject in depth as related to the quality control of fruits and vegetables. The reader is referred to these reviews for more detail than can be given in the context of this chapter.

2. General Considerations

2.1. BASIC ELEMENTS

Most of the instrumental methods of texture description are based on mechanical tests which involve measuring the resistance of the food to applied forces greater than gravity. In general, all the devices used for that purpose consist of four basic elements:

(a) *a probe* contacting the food sample; this may be a flat plunger, a pair of shearing jaws, a tooth-shaped attachment, a piercing rod, a penetrating cone, a spindle, a cutting blade, or a set of cutting wires;

(b) *a driving mechanism* for imparting motion (vertical, horizontal, rotational or levered) to the probe at either a constant or a variable rate; it may vary from a simple weight and pulley arrangement to a more sophisticated hydraulic system or a variable drive electric motor;

(c) *a sensing element* for detecting the resistance of the foodstuff to the applied force (cutting, piercing, compressing, shearing, etc., depending on the probe and the type of motion); it may be a simple spring or a more sophisticated strain gage transducer;

(d) *a read-out system* which may be a maximum force dial, an oscilloscope, a recorder tracing the force-distance or the force-time curve, or an electronic integrator which registers the amount of expended energy.

Most, but not all, instruments use a flat platform for supporting the solid specimens and a cup-like container for liquid or semi-liquid products.

There are a few consistency measuring devices which involve gravity forces only. These, exemplified by spreadmeters and flowmeters, are based on different principles. There is no probe in contact with the specimen and no sensing element. The driving force is supplied by the weight of the material being tested and the read-out system usually consists of a graded scale (either linear or concentric) which allows the degree of spread to be determined after a specific period of time.

2.2. DESTRUCTIVE VS NON-DESTRUCTIVE TECHNIQUES

Most mechanical texture measurements are destructive in that the applied force exceeds the strength of the test food which disintegrates in the process. This precludes making several tests on the same sample. Because of this, non-destructive methods have recently come under consideration (e.g. Abbott *et al.*, 1968a, b). These, however, pose specific problems when attempts are made to correlate them with the sensory evaluation in the mouth which is destructive by its very nature. They will be discussed in more detail in a subsequent section.

2.3. CLASSIFICATION

Instrumental methods of texture measurement may be classified into groups according to several schemes. Scott Blair (1958) recognized three categories based on the general nature of the test: fundamental, empirical, and imitative. Fundamental tests measure fundamental rheological properties (e.g. viscosity and elasticity) and aim at relating the nature of the tested product to basic rheological models. Empirical tests measure parameters, often poorly defined, that practical experience indicates are related to textural quality. Most popular devices (e.g. penetrometers, shearing instruments) fall in that category. Imitative tests imitate conditions to which the material is subjected in practice (e.g. biting, spreading, kneading).

Drake (1961 and 1966) proposed a classification system based on the motion and geometry of the test specimen. The general scheme included the following types of motions: rectilinear (parallel, divergent, convergent), circular (rotation, torsion), axially symmetrical (unlimited, limited), other defined (bending, flexural), undefined (mechanical 'treatment').

Bourne (1966a) developed a classification based on variables underlying the measurement and divided the various methods into force, distance (including area and volume), time, energy, or ratio-measuring, multiple-measuring, and multiple-variables. This is illustrated in Table I. Other researchers used a commodity classification (e.g. Finney, 1969a) or grouped the instruments according to the general type of exerted force (e.g. Kramer and Twigg, 1970; Gordon, 1967). The latter, with some modification, will be the most familiar to practicing food technologists interested in practical applications of texture measurements.

3. Types of Texture Measuring Devices

3.1. PENETROMETERS

The penetrometers represent one of the oldest and the most time-honored groups of texture measuring instruments. As the name implies, they are based on the principle of penetrating the test material, usually with a tapered rod-like or a cone-shaped probe, and measuring either the force required for a given penetration depth or the total depth of penetration. The former method is more popular. The higher the force reading (or the smaller the penetration depth), the more resistant is the material. This is taken as a measure of 'firmness' or 'toughness'. Some penetrometers use one probe,

TABLE I

Classification of texture testing devices according to the measured variable
(after Bourne, 1966a)

Measured variable	Dimensions	Typical devices
force	$ml\,t^{-2}$	Magness-Taylor Pressure Tester Tenderometer Maturometer Warner-Bratzler Shear Kramer Shear Press
distance	l	Ridgelimeter Ball Compressor Bostwick Consistometer
area	l^2	Adams Consistometer
volume	l^3	Succulometer
time	t	Ostwald Viscometer
energy	ml^2t^{-2}	Farinograph Meat Grinder
ratio	dimensionless	General Foods Texturometer
multiple-variables (controlled)		MIT Denture Tenderometer General Foods Texturometer
multiple-variables (uncontrolled)		Durometer

while others use several probes mounted on a plate. A review of the various types of penetrometers was published by Bourne (1966b).

Fruit pressure testers are one of the oldest instruments of that type. The first design should be credited to Morris who in 1917 constructed a simple device consisting of a marble embedded in paraffin and resting on a spring scale (Morris, 1925). The marble was pressed against the fruit until it penetrated as far as the paraffin and the exerted force was noted. In 1925, Magness and Taylor described a pressure tester which measured the maximum force needed to press a 5/16" diameter metal tip into the fruit. This device became known as the Magness Taylor fruit pressure tester and has been widely adopted for judging fruit maturity. Different variations on the basic design and their practical applications were reviewed by Haller (1941). A special attachment intended to simulate the action of the human thumb pressing against the fruit without penetration was developed by Schomer and Olsen (1962). Called the 'mechanical thumb', it is shown in Figure 1. It consists of a probe and a light assembly connected with a wire. The probe is a 1/2" diameter segment of a 1" sphere; it is pressed into the fruit a pre-set distance of approximately 0.055" until the lamp in the light assembly turns on. The force is then read from the scale.

The major drawback of these devices is that the force is applied manually, thus leading to a poor control of the rate of application. In addition, Bourne working with apples has shown (1965) that in the case of the Magness Taylor pressure tester the measured force can be equal to or greater than the bio-yield point depending on the

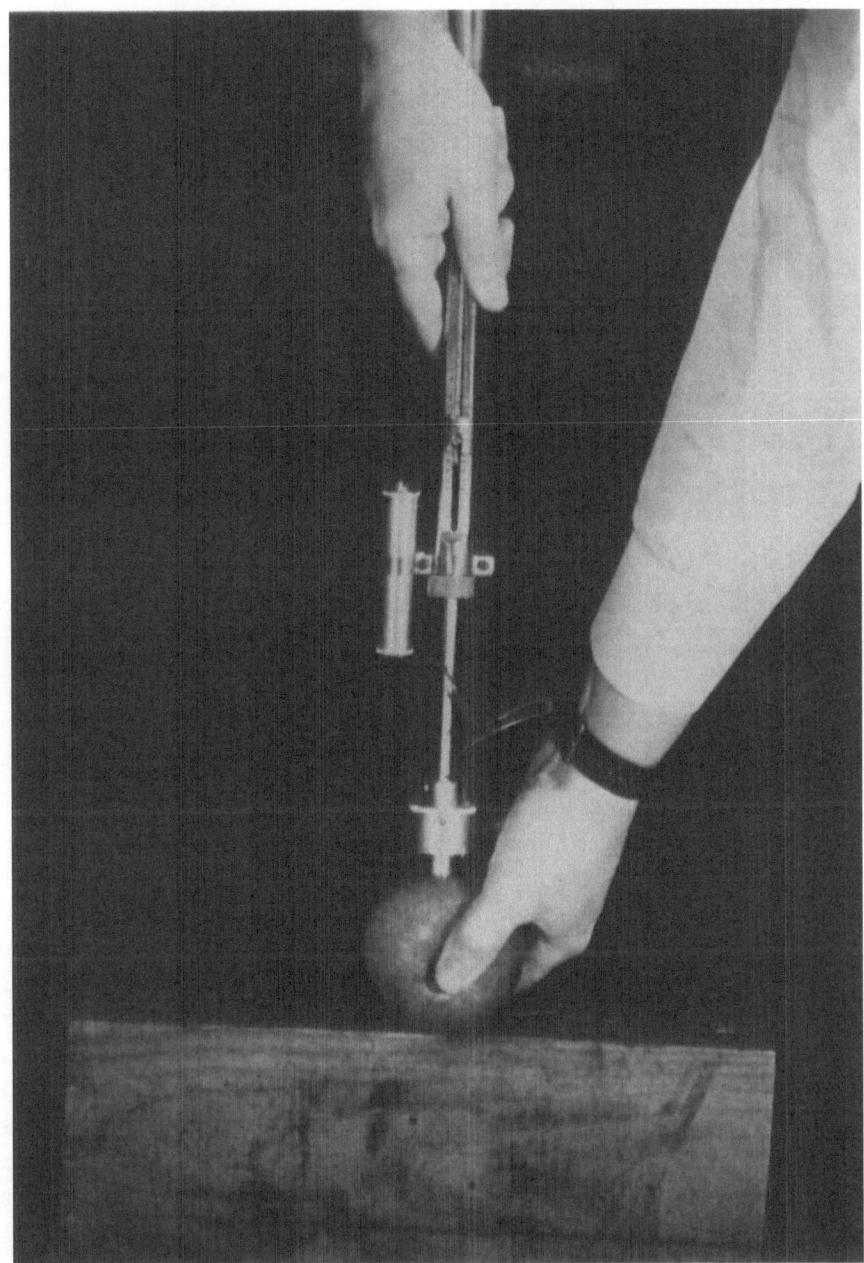

Fig. 1. Magness-Taylor pressure tester (USDA photograph BN 18678).

type of fruit tested (Figure 2). He suggested that the depth of penetration be greatly reduced to obtain a consistent measurement of the yield point in a manner correlatable with the thumb test. The 'mechanical thumb' modification of the pressure tester incorporates that feature. The dimensions of the probe also influence the results.

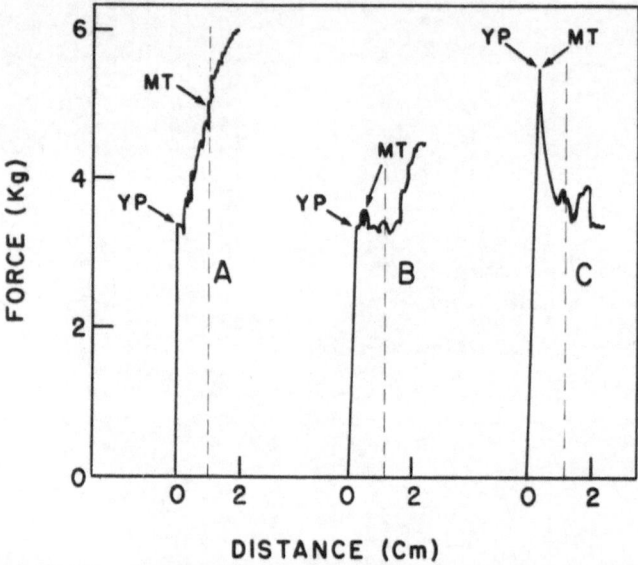

Fig. 2.   Force-distance curves for penetration tests on apples (5/16″ diam. tip), Bourne, 1965. –
YP = yield point; MT = reading which would have been obtained from the Magness-Taylor fruit
pressure tester.

Factors inherent in the tested fruit sample that affect the reading were summarized by
Gordon (1967).

Examples of penetrometers utilizing more than one probe are the Christel Texturemeter (Christel, 1938) and the Maturometer (Mitchell *et al.*, 1961). The Texturemeter
uses 25 punches approximately 3/16″ in diameter, while the Maturometer employs
143 probes 1/8″ in diameter (Figure 3). Both of these instruments have been originally
designed for testing peas. In the Texturemeter, the sample is held in a box into which
the punch assembly is pushed. In the Maturometer, a number of peas equal to that
of the penetrating probes are placed in the countersink of 3/16″ diameter holes
located directly below the punch assembly. Each probe penetrates one pea and the
resulting force reading represents a composite for the lot. Thus, sample size is selected
numerically for the Maturometer and by volume for the Texturemeter. More recently,
Casimir *et al.* (1971) fitted a single Maturometer probe into the Instron Universal
Testing Machine in order to be able to detect textural differences caused by genetic
and environmental factors in a situation where sample size was limited.

Both of these multiple-probe penetrometers have also been applied to the testing of
meat tenderness. Using a similar approach, Armour and Co. (Hansen, 1972) designed
an instrument where the force required to insert manually 10 pointed probes is detected
by strain gages and used to classify meat carcasses.

Penetrometers have also been used for testing the rigidity of gels. The best known
is the Bloom gelometer (Bloom, 1925), a standard instrument for measuring the
gelling power of edible gelatin. With this device, the rigidity of the gel is measured as
the weight of lead shot required to push a standard plunger 4 mm deep into the gel. A

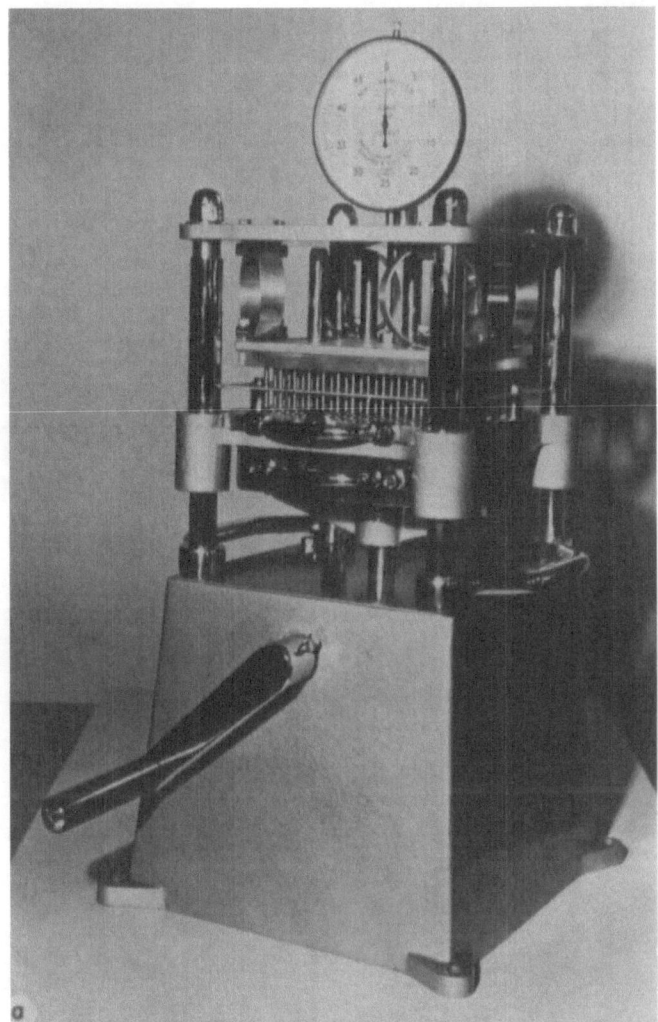

Fig. 3.   Lynch-Mitchell Maturometer (courtesy of L. J. Lynch, CSIRO, Australia).

transistorized version of the instrument has been described (Schachat and Nacci, 1960).

An example of devices based on depth measurement are cone penetrometers, representative of which is the ASTM grease penetrometer (Figure 4). This instrument measures firmness of unctuous materials in terms of depth to which a metal cone, of standard size and shape, sinks into the surface of the sample under defined conditions. Haighton (1959) discussed the rheological principles of cone penetrometer measurements on fats and showed how yield values can be calculated from the obtained penetration data. He listed the following factors as affecting the yield value: smoothness of the cone, sharpness of the cone tip, structural hardness and work hardening (or softening) of the test material, duration of penetration and kinetic energy of the cone.

The consistency of mayonnaise and similar semi-solids has been determined ob-

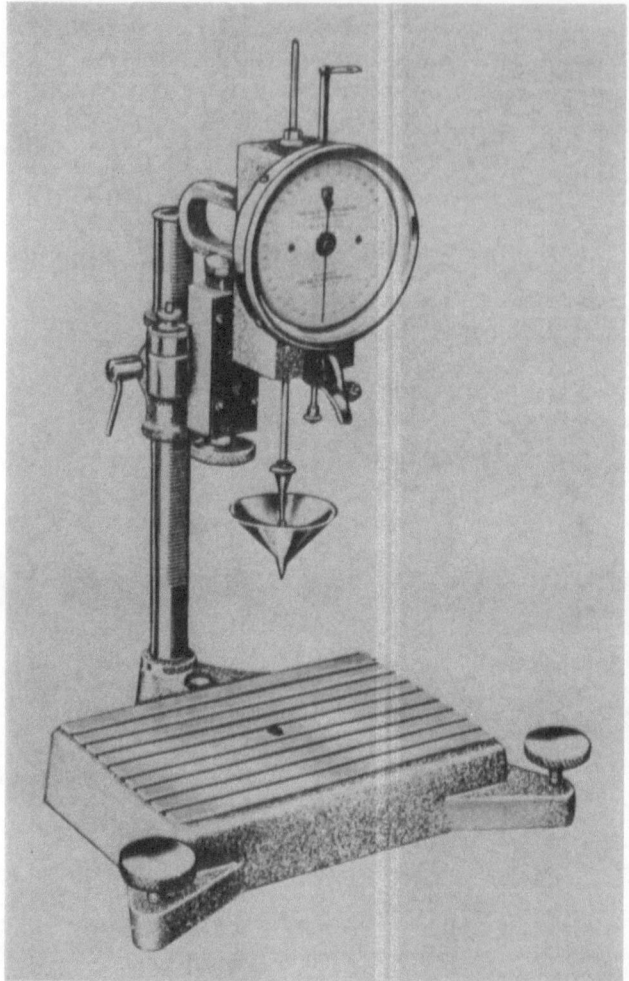

Fig. 4.   ASTM grease cone penetrometer (courtesy of Precision Scientific Co.).

jectively by dropping a 'plumit' (a pointed metal rod with engraved scale divisions) from a fixed height and reading the depth of penetration.

Several workers have recently fitted various penetrometer probes into the Instron Universal Testing Machine or similar instruments. This offers the advantage of controlling the rate of force application and, assuming a recorder is used, of recording both force and distance of penetration. Such an arrangement provides not only a more accurate determination but also permits one to study the effect of variables on test results. Owing primarily to the work of Professor Bourne at the New York State Agricultural Experiment Station, Cornell University, and to that of Professor de Man at the University of Guelph, Ontario, a better understanding has been obtained recently of what is actually being measured in penetration tests.

Using flat probes, Bourne (1966b) showed that penetration tests of cellular plant material involve both shear and compression. In addition, flow appears to take place when testing materials containing substantial amounts of fat (deMan, 1969). The puncture force can be expressed as:

$$F = K_s P + K_c A + K_f A + C$$

where F=force measured; K_s=shear coefficient of the food; K_c=compression co-efficient of the food; K_f= flow coefficient of the food; P=punch perimeter; A= punch area; and C=constant.

This indicates that the area and the perimeter of the probe would have a definite influence on force readings and that this influence will vary depending on the rheological properties of the test material. Tapered probes pose a very complicated situation because both the perimeter and the area change with the depth of penetration.

Tanaka et al. (1971) discussed the various attempts to relate measurements by cone penetrometers to fundamental physical properties and reported on their own results dealing with the characteristics of the force required to push a cone into a test food at a constant rate. The resulting deformation is related to both the plastic and the viscous nature of the food and the resulting stress can be expressed as:

$$F'/A' = \eta_{app}(dh/dt) + f = F \cot(\theta/2) \cos(\theta/2)/\pi h^2$$

where F^1=actual force applied to the sample; A=deformation arc; h=depth of penetration at time t; θ=angle of the cone; f=yield value of the food; η_{app}=apparent viscosity of the food; and F=force applied with the cone.

Thus, the registered force is a composite of both the yield value and the apparent viscosity of the test material. The two can be separated by plotting the penetration speed (dh/dt) on the abscissa and the penetration stress $(F \cot(\theta/2) \cos(\theta/2)/\pi h^2)$ on the ordinate. The slope of the plot gives the apparent viscosity and the intercept at zero speed gives the yield value.

3.2. COMPRESSIMETERS

As the name implies, compressimeters test the resistance of the food to compression. Similarly to penetration test, this can be measured either as force needed to produce a given deformation or as deformation caused by a given force. Compressimeters differ from the penetrometers in that the test material is not pierced or punctured and, usually, the yield point is not exceeded.

These devices can be divided into several groups depending on the type of surface-to-surface contact: a flat plunger compressing a flat specimen, a curved plunger compressing a flat specimen, or a flat plunger compressing a curved specimen. Combinations of a spherical plunger and a spherical specimen surface are rare.

For many years, the most popular application of compressimeters has been in measuring bread softness. This was prompted by much attention being paid to the problem of staling, resulting in increased firmness. The amount of compression under constant load was taken as an index of 'softness' and the force required to cause a given deformation was taken as an index of 'firmness' (Babb, 1965).

Probably the best known instrument in this category is the Baker compressimeter, a standard Cereal Laboratory Method. It consists of a flat plunger that depresses a slice of bread by rotation of a small drum actuated slowly and uniformly by a motor and shaft beneath the apparatus. The force acting on the plunger and the amount of compression can be read from two scales. Platt and Powers (1940) reported the results of detailed studies dealing with the effects of bread slice thickness, amount of stress applied, rate of deformation, etc.

By using more refined designs, a recorder and strain gages to detect forces of compression, both force and distance variables can be measured in an accurate and continuous manner. These features were used by Babb (1965) in designing a more modern instrument for testing bakery goods, and by MacAllister and Reichenwallner (1959) in constructing an apparatus for gel characterization. Many researchers mount flat plungers into the Instron or other similar devices.

Care must be exercized that the plunger does not cut into the sample and, thus, becloud compressive with shearing forces. Whenever possible, the plunger should be larger than the sample and the test specimen should have an even, flat surface so that the area in contact with the plunger is constant and known. For comparative purposes, specimens of identical area and height should be used. For more fundamental studies, the results should be expressed as the stress/strain (modulus of elasticity) taking into account specimen dimensions. Differences in sample diameter can often lead to serious errors as larger pieces tend to deform more (Bourne, 1967).

The above errors can be avoided if the test is conducted in such a way that true compression (volume contraction) is the only type of deformation that the specimen undergoes. This can be done by shaping the test material to fit snugly into a container and compressing it with a plunger of such dimensions that there is little or no clearance between the plunger and the sides of the container (see e.g. Kramer, 1961). In such tests, care must be exercized not to confound the results with the resistance of the entrapped air.

Compressive loading between two flat plates has received both theoretical and practical considerations from researchers following the engineering approach to food texture measurements. Hammerle and McClure (1971) stated that the two main difficulties of the compression test are lateral instability and end restraint, both of which result in nonuniform stress distribution. From the practical standpoint, these effects can be minimized by using relatively thin specimens and relatively low compressive forces. The latter has also been shown to give a better resolution between similar samples (Bourne, 1967) and to be more similar to the consumer evaluation of firmness by the hand 'squeeze' test.

An example of a compressimeter using a hemispherical probe is the ball compressor devised by Caffyn and Baron (1947) for measuring firmness of large heads of cheese. A load is applied to the probe resting on the specimen by means of a series of lever arms and the cheese deformation is measured before and after the removal of the load. The first reading denotes sample firmness while the difference between the two readings is taken as a measure of elastic recovery. Applications of this device to the description of a number of commercial cheeses has been discussed by Cox and Baron (1955). A similar instrument has been constructed for testing meat (Palmer, 1962).

Compression of a curved specimen with a flat plunger is exemplified by a test for firmness of onions developed by Ang *et al.* (1960) using a square plunger fitted into the Kramer Shear Press and by various compression devices (e.g. Voisey and Hunt, 1967) used to test firmness or breaking stress of egg shells. Typical of the latter is the application of the pneumatic loading unit devised by Mohsenin (1963) to deforming eggs between two flat plates at a controlled rate and determining the force with a strain gage load transducer mounted beneath the lower platform (Richards and Staley, 1967). A recent, more sophisticated approach to egg testing takes into account the shape of the specimen and the uneven distribution of stresses (e.g. Hammerle, 1969).

The Hamden deformation tester developed by Rowlands (1964) for measuring firmness of Brussels sprouts is an interesting example of a case where a curved specimen is compressed with a concave plunger. This plunger shape was necessitated by the fact that a flat-surfaced plunger caused some damage to the outer leaves, thus making it impossible to measure sprout compression alone.

A very unique compressimeter is represented by the Firm-o-meter designed by Kattan (1957) for measuring firmness of tomato fruit. In this device, the fruit is compressed around its entire perimeter by means of a link chain. One end of the chain is firmly attached to the framework while the other passes over a sprocket, encircles the fruit, fits over a second ballbearing sprocket and is then attached to a pulley and weight arrangement. The compression of the tomato is determined as the differences in the deflection on the pointer attached to the pulley between 500 and 2000 g on the weight pan. The first weight is used to merely tighten the chain.

A similar idea of compressing the specimen around the entire perimeter was recently applied by Howard and Heinz (1970) to measuring the firmness of carrots by means of a special test cell fitted into the Instron.

3.3. SHEARING DEVICES

This section will be limited to shearing devices used to test solid foods. Instruments based on the principle of shear and used for the characterization of liquids and semi-solids will be covered in the section dealing with viscometers.

Shear apparatuses usually employ a single blade or a multiple-blade probe. Their most popular use has been in meat texture work where force required to shear a specimen has been taken as a quantitative measure of tenderness. Much dissatisfaction has been expressed with the method, and the controversy continues concerning what is tenderness and whether or not shear measurements are a true reflection of sensory meat tenderness. More recently, the use of shear devices has spread to vegetables, fruits, cakes, and other products.

An example of a single blade shear apparatus is the well-known Warner-Bratzler Shear (Figure 5). Originally designed by Warner (1927) and subsequently modified by Bratzler (1932), the unit is composed of a 1 mm thick metal blade with a triangular hole into which is inserted a cylindrical sample of meat. The blade is led through a slit between two shear bars which move downwards and shear the sample. The maximum force exerted on the blade is detected by a dynamometer spring and read from a dial scale. Much work has been done with this device regarding its performance, correlation with other objective tests, and with sensory evaluation (Szczesniak and

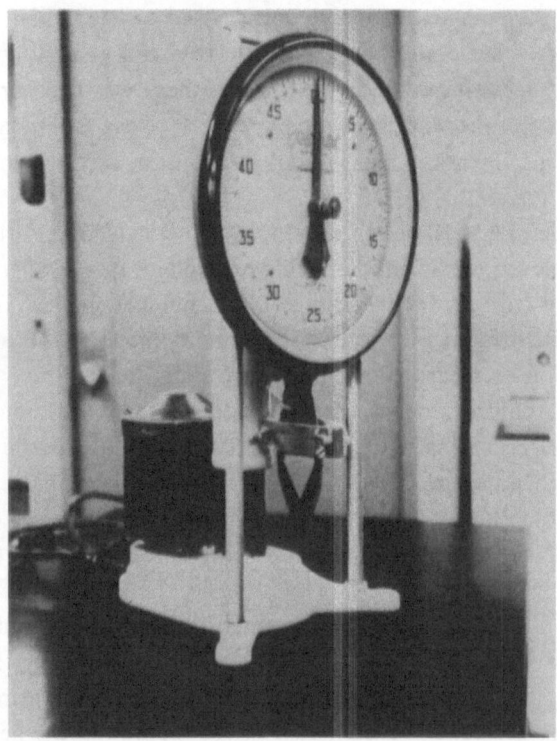

Fig. 5.   Warner-Bratzler Shear.

Torgeson, 1965). Two recent modifications involving the use of more sensitive means of force detection and a recording system were published by Spencer *et al.* (1962) and Voisey and Hansen (1967).

The Warner-Bratzler Shear has been used extensively for testing meat, but its reliability has often been questioned. Very recently, two groups working on different types of meat have shown independently that small variations in sample diameter have a very large influence on force readings (Pool and Klose, 1969, and Davey and Gilbert, 1969). A mechanized method of sample preparation, which should give a much more precise control over sample diameter than is possible with the manual use of a cork borer, has been described by Kastner and Henrickson (1970).

Examples of multiple-blade shearing devices are the Pea Tenderometer, the Kramer Shear Press, the Pabst Texture Tester, and Dassow's Shear-jaw device.

The Pea Tenderometer (Martin, 1937) consists of two grids through which the peas (or other product) are sheared, motive power for moving one grid with respect to the other at a constant rate, and a pendulum mechanism for measuring the exerted force. In spite of many criticisms, the apparatus continues to be used widely as an instrument for grading tenderness of peas and for establishing the bases on which farmers are paid for their crop.

The Kramer Shear Press (Kramer, 1961) uses a stationary rectangular box with slots in the bottom to hold the sample and a moving probe composed of 10 bars, 0.114″ thick, to drive through the test specimen (Figure 6). The bars are moved by a piston driven at a selected speed, with the hydraulic drive system powered by a gear pump attached to an electric motor. The resistance of the sample is measured by the compression of a proving ring and is read in pounds force either from a constant dial read-out system or from a force-distance recorder. A model is available in which the results are read off directly in Tenderometer units for easy comparison with the Pea Tenderometer.

The Pabst Texture Tester (Figure 7) is a shear device in which the motion of the shearing blades is horizontal rather than vertical and the cell is very small, being

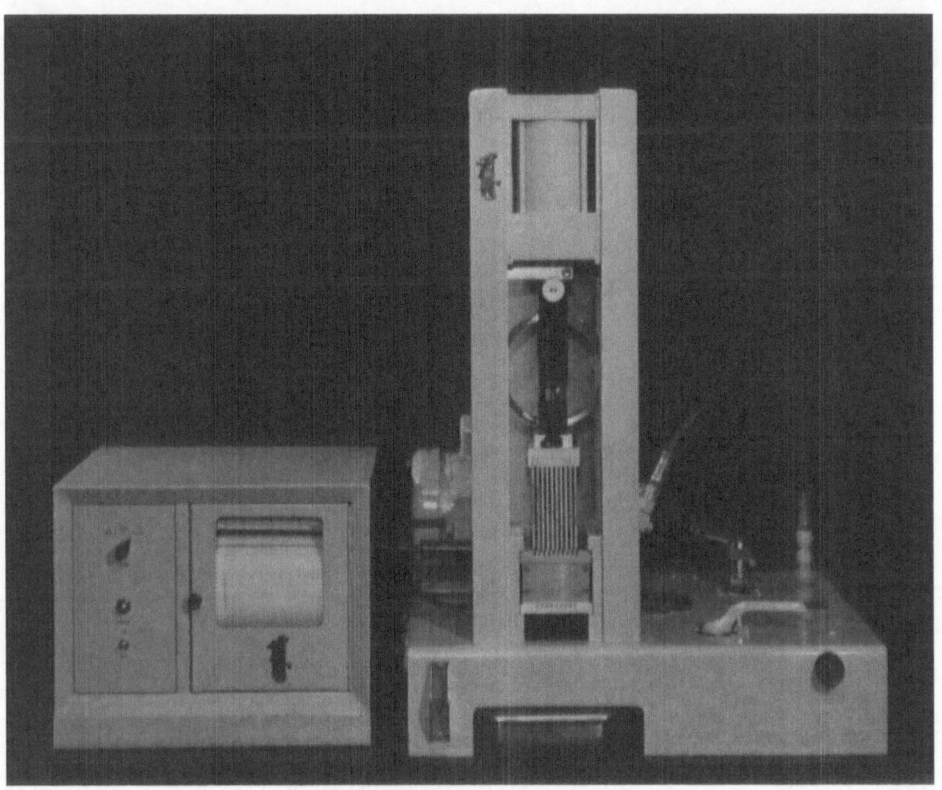

Fig. 6a.   Kramer Shear Press, basic unit; presently known as the Texture-Test System.

Fig. 6b. Various cell types for use with the Texture-Test System; left to right, top row – standard shear-compression cell, combination single blade and meat shear cell; bottom row – universal cell with a tight fitting plunger (also shown is the back extrusion plunger and different removable cell bottoms), succulometer cell.

designed to hold bite-size pieces. The instrument is powered by hydraulic pressure and is a self-cleaning, automatic device aimed at quality control applications. Because individual peas, beans, corn flakes, etc., can be tested, the apparatus permits the determination of the distribution pattern, as well as of an average resistance for a lot of the test material (Szczesniak, 1968).

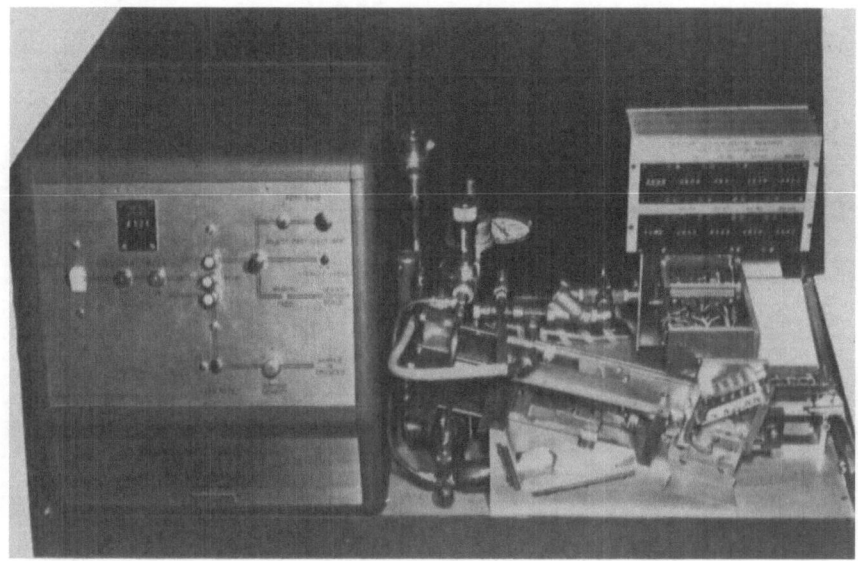

Fig. 7. Pabst Texture Tester.

Dassow's shear-jaw device is interesting in that it employs a levered action (Dassow *et al.*, 1962). A modification of an earlier device, the instrument consists of a hydraulic system and a working platform composed of a moving multiplated shear jaw and a stationary multigrid plate (Figure 8). The movement of the jaw shears the material placed on top of the stationary plate. The device was designed for work with fish and shellfish, and its use has not been extended to other products.

When a recording type instrument is used, both maximum force readings and areas under the peak (work function) can be determined. Some workers prefer maximum force, while others prefer area measurements. In general, the former is used more often since it is easier to determine and since no definite advantages of area measurements have been demonstrated.

All shearing devices are subject to error caused by differences in sharpness of the blades, alignment, warp, etc. Difficulties in standardizing the entire unit including the cell have not been resolved. In addition, questions are being asked as to what exactly is being measured by such instruments.

It appears that, similarly to penetration tests, a composite of different forces is at play. Using the above described Kramer Shear Press on a large variety of different food types, Szczesniak *et al.* (1970) found that the measurement involves shear,

Fig. 8.   Dassow's hydraulic shear-jaw device (courtesy of J. A. Dassow, Bureau of Commercial Fisheries, Seattle, Wash.)

compression, and extrusion, and that different combinations of these forces may act on a given food depending on its rheological nature. It is probable that a similar situation exists with respect to other shearing devices.

The same study also reported on the effect of sample weight on maximum force and peak area readings. The relationship between maximum force values and sample weight was found to be different for different foods. Products could be grouped into three general categories: those exhibiting a constant force to weight ratio, those exhibiting a continuously decreasing force to weight ratio, and those exhibiting a constant force, independent of sample weight, beyond a certain cell fill level (Figure 9). Since most foods exhibited a nonlinear relationship between maximum force and sample weight, Shear Press data should not be expressed as units of force per unit of sample weight unless it can be shown that the relationship is linear in the region of interest. Peak areas showed an exponential relationship to sample weight according to the equation $A = 10^a W^b$, where the exponents 'a' and 'b' appear to be characteristic of the amount and the general type of resistance offered by the food.

3.4. CUTTING DEVICES

These units usually employ a knife-like blade or a wire to cut through the sample.

They have been used mainly to test meat, cheese curd, and fiber containing vegetables such as asparagus, although some researchers have also applied them to testing 'cutting firmness' of products such as tomatoes and pineapple.

An early instrument of this type is the asparagus fiberometer developed by Wilder in 1947 for the National Canners Association. It consists of a supporting block of a parabolic shape with vertical slots 0.039–0.042″ wide through which the cutting wire is passed to press at right angles against the asparagus. The length of the asparagus that can be cut with 3 lbs. pressing on the wires is taken as a measure of fibrousness. The shorter the length, the tougher and more fibrous is the sample (Townsend *et al.*, 1956).

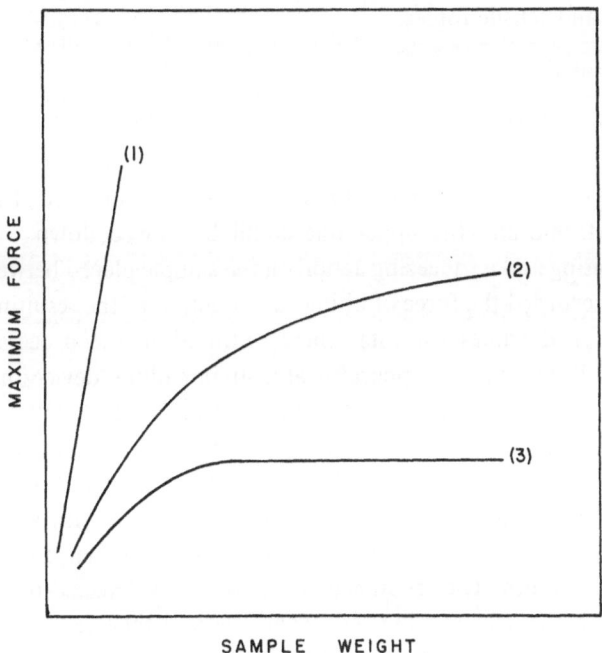

SAMPLE   WEIGHT

Fig. 9.   Effect of sample weight on maximum force readings with the standard shear-compression cell (Szczesniak *et al.*, 1970). Behavior (1) is exemplified by white bread and sponge cake, (2) by raw apples and cooked white beans, and (3) by canned beats, canned and frozen peas.

Kramer *et al.* (1949) described two devices for measuring fibrousness in asparagus. Both are an adaptation of the fruit pressure tester and use a 0.017″ thick stainless steel knife fitted to a shaft to cut through a stalk. One device uses a dial gage and the other a scale to measure the force required to sever the sample.

The Cherry Burrell curd tension meter is an example of a cutting device used for testing cheese curd or soft cheeses. A round frame containing radially oriented cutting blades is lowered into the sample-holding container and the force required to cut through the specimen is registered either on a dietetic scale beneath the container or, in later models, by the blades shaft. A modification using a strain-gaged cantilever beam to support the sample container and to record the force has been described by Voisey and Emmons (1966).

A rotating knife tenderometer for quantifying meat tenderness was described by Bjorksten *et al.* (1967). The device utilizes a mechanized cup-shaped knife which rotates at a constant speed for a predetermined number of revolutions per test. The depth of cut or the total area of the cut surface is said to be proportional to sample tenderness.

Even more than the shearing devices, these instruments are subject to errors due to the sharpness of the knife and the speed with which the force is applied. Although no specific studies have been published on the type of forces involved in the 'cutting' process, it is suspected that – similarly to the 'shearing' process – they include compression, shear, and tensile forces.

3.5. Masticometers

The Volodkevich (1938) bite tenderometer was probably the first of a series of attempts to measure the textural properties of foods under conditions approaching those of mastication. Two wedges with rounded points substituted for teeth. The lower wedge was fixed on a frame and the upper one could be moved down with two levers, thus exerting a biting and a squeezing action on the sample placed between the wedges. The instrument recorded the force of biting as a function of the resulting deformation and could also yield values for total energy utilized in the process. The various modifications of Volodkevich's apparatus and similar biting devices used for testing meat have been described in detail by Szczesniak and Torgeson (1965).

The next significant development in this area came in the form of the MIT Denture Tenderometer (Proctor *et al.*, 1955, 1956a). This instrument utilized a complete set of human dentures fitted into the Hanau articulator driven with an electric motor. This provided for a mechanical chewing arrangement similar to the motions of mastication in the mouth. The resistance of the food was detected by a pair of strain gages mounted on the arm moving the upper jaw and then fed into a cathode-ray oscilloscope. A permanent record was obtained by photographing traces on the face of the oscilloscope with a Polaroid camera. Probably the best description of the various modifications of this apparatus and ways of using it for testing different foods can be found in the Ph.D. thesis by A. Brody (M.I.T., 1957). An adaptation of this instrument to the testing of peas and correlations with the Pea Tenderometer have been described (Proctor *et al.*, 1956b).

A simplified design combined with a more advanced interpretation of the recorded force patterns is portrayed by the General Foods Texturometer (Figure 10). Described by Friedman *et al.* (1963), this instrument replaced the oscilloscope with a strip chart recorder and the dentures with a plunger and plate arrangement. The sensing element was moved from the articulator arm to the sample area in order to eliminate gravity effects. The recorded force-time curves represent a permanent record of the textural spectrum of the test material (Figure 11). Textural parameters (Szczesniak, 1963b) of hardness, cohesiveness, springiness, fracturability, and adhesiveness can

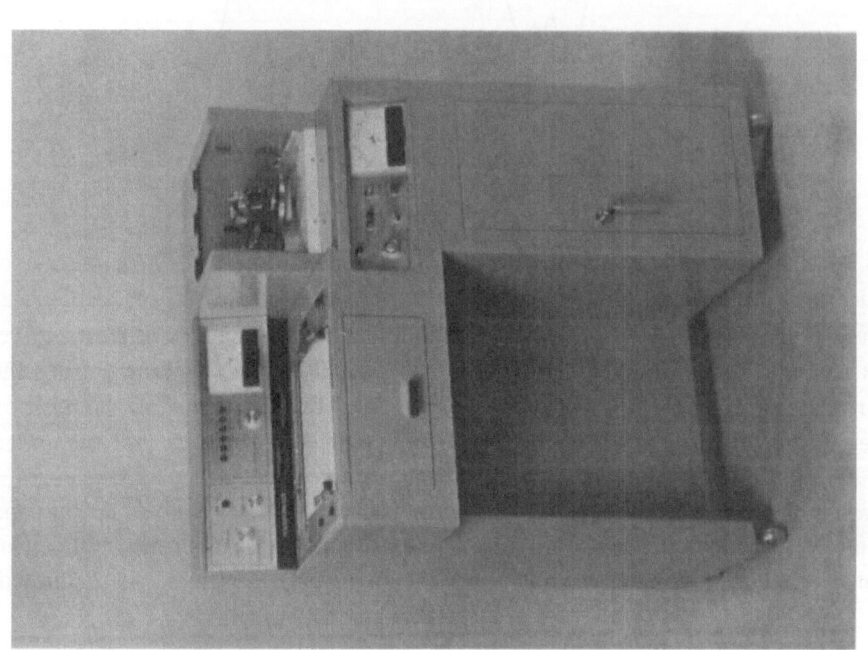

Fig. 10. Commercial model of the GF Texturometer, left – the entire unit, right – a close-up of the 'masticator'.

be read off the curves, and parameters of chewiness and gumminess can be calculated. Drake (1966) added the 'index of crushability' to the interpretation of Texturometer curves. The instrument quantifies several textural characteristics in one test and can, therefore, be classified as a controlled multiple variables measuring device. It is adaptable to testing a variety of solid and semi-solid foods for which plungers of different materials, sizes, and shapes can be employed to more closely approximate the type of action which a food undergoes in the mouth (e.g. biting with front teeth, crushing with molar teeth, compressing between the tongue and the palate, etc.). Because, similarly to the human jaw, the plunger moves in a levered motion with

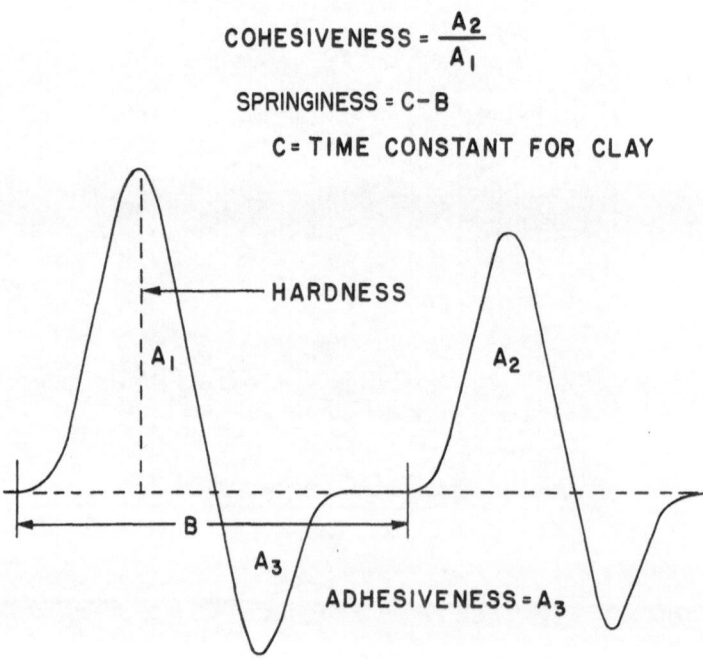

Fig. 11.   Typical Texturometer curve.

speed as a sinusoidal function of time, the abscissa is only an approximation of distance and a correction factor is needed for accurate force/distance calculations. Kapsalis (1967) reported on a graphical method for converting force-time to force-displacement curves. Another complication is the downward bending during the test of the sample supporting platform. This is caused by the nature and the position of the force-detecting strain gage. This can, however, be calculated and accounted for if an exact calculation of penetration distances is desired.

Because the interpretation of Texturometer curves is based on a classification system which already attempts to interpret composite mechanical parameters in terms of sensory texture attributes, correlation with an organoleptic panel is facilitated.

As is to be expected, the quality of such correlations tends to depend on the type of panel used and, to a certain extent, on the tested foodstuff.

Excellent correlations were reported (Szczesniak *et al.*, 1963a) for a variety of foods with organoleptic evaluations by a trained texture profile panel (Brandt *et al.*, 1963). Working with a less expert panel, Brennan *et al.* (1970) found a good correlation only for the parameter of hardness which is the easiest for a panel to quantify. The same group also reported on the importance of controlling the sample size and demonstrated a linear positive relationship between hardness values and both horizontal and vertical sample dimensions. Other parameters were not affected by sample size. This indicates that, similarly to other texture testing devices, an accurate sample preparation and careful control of sample size are essential to the obtaining of satisfactory results. Kapsalis *et al.* (1970) used a modified Texturometer on precooked freeze-dried beef of different moisture contents and obtained high correlations between instrumental values for hardness, cohesiveness and crushability index, and sensory attributes of ease of biting and tenderness.

Of special appeal to researchers interested in simulating the chewing motion of the human mouth is a device recently described by Molnar (1968) which simulates the motion of the muscles, as well as that of the teeth.

3.6. CONSISTOMETERS

The name 'consistometer' has been given to various types of empirical instruments used for testing semi-solid foods; basically, however, none of these devices measure consistency in the rheological sense. According to Reiner and Scott Blair (1967) consistency is "the property of a material by which it resists permanent change of shape, defined by the complete stress-flow curve."

Empirical consistometers are based on a variety of principles but most of them fall into two general sub-classes: devices which measure distance of spread and devices which measure resistance to a rotating spindle or paddle.

Typical of the first group are the Adams and the Bostwick Consistometers. The Adams Consistometer measures the area to which a given quantity of the test material will spread under a certain set of conditions. The Bostwick Consistometer measures the distance a given amount of the semi-solid will travel down a slanted trough upon being released from a container. Both devices are limited to materials which are fluid enough to flow under the test conditions. The Bostwick Consistometer is an official National Canners Association test method for the consistency of catsup. As such, it has also been widely used with other concentrated tomato products, e.g. paste, sauce, puree. In many cases, however, tomato pastes of high solids content (e.g. 30–35%) will not flow in the Bostwick Consistometer and recourse must be made to other test methods.

The Adams Consistometer has been designed originally for use with cream-style corn. It consists of a large metal or plastic disc upon which are engraved 20 concentric circles, 1/4" apart, and a truncated steel cone which fits tightly against the disk so that its circumference coincides with the smallest, innermost circle. The cone is pushed tightly against the disk, filled with the test sample, and lifted vertically up-ward allowing the material to flow over the disk in an unrestrained manner. The extent of flow is determined after 30 s by averaging the distance of spread at four quadrant points. This type of a consistometer has also been used on fruit preserves, salad dressings, pancake batter, pie fillings, farinaceous pastes, etc. It is quite obvious that measurements by such devices represent a composite of several physical para-meters including surface tension and friction.

Consistometers which measure the resistance to a rotating element include the well-known Brabender Amylograph which measures the apparent viscous behavior of farinaceous pastes when subjected to heating and cooling. Although it may be visualized as imitating the effect of cooking or baking, the Amylograph is similar in practice to a viscometer. It consists of a cup revolving at a constant speed and surrounded by an air bath for temperature control. Viscosity is determined as the torque on the measuring unit composed of a disc with several short rods extending into the sample. Because the instrumental constant is not known, the results are expressed in arbitrary 'Brabender Units' rather than in fundamental units. The instrument is equipped with a revolving drum recorder which gives the disadvantage of non-rectangular chart axes. Voisey (1971) has shown how rotating consistometers of this type can be converted into more accurate, recording-type instruments by using electronic transducers.

Typical of the consistometers falling outside the two categories described above is the Consistometer for measuring spreadability and hardness of butter. Originally developed by Huebner and Thomsen (1957) and subsequently modified by Kapsalis et al. (1960), the instrument consists of a pendulum driven by a constant speed electric motor. Spreadability is measured as resistance to the spreading action of a knife, and hardness is measured as resistance to cutting by a wire. The knife or the wire is mounted at the lowest point of the arc described by the pendulum and the total resistance of the test sample is measured by the sum of forces exerted by fixed weights acting through a pulley system and the constant speed motor. The former is measured directly in grams and the latter is registered on a torque-meter mounted on the pendulum arm.

3.7. Viscometers

An excellent treatment of fundamentals of viscosity measurements and descriptions of different types of viscometers can be found in the book, *Viscosity and Flow Measure-ment* by Van Wazer *et al.* (1966). Only some general comments related to the type of information generated by viscometers commonly used for testing foods will be given here.

The two types of viscometers most common to a food laboratory are rotational and capillary viscometers. Typical of the first class are the MacMichael and the Brookfield Viscometers. The MacMichael Viscometer consists of a cup and an inner cylinder suspended from a wire. The test material is placed in the cup, and the inner cylinder is immersed in it. The cup is rotated and the torsional force on the wire required to hold the inner cylinder stationary is taken as the measure of the consistency of the test material. The Brookfield Viscometer uses the reverse situation. The container holding the test material is held stationary and a spindle is rotated inside it. Spindles may have a variety of shapes and sizes, depending on the characteristics of the test material. A number of different models of the Brookfield Viscometer are available.

Since most food substances are non-Newtonian, the rotating speed has a significant effect on the viscosity of the material. It is generally recognized that one-point viscosity measurements are of very limited value (and can often be very misleading) unless the test material exhibits Newtonian flow (Figure 12). During sensory evaluation, the specimen is subjected to different rates of shear due to manipulation by

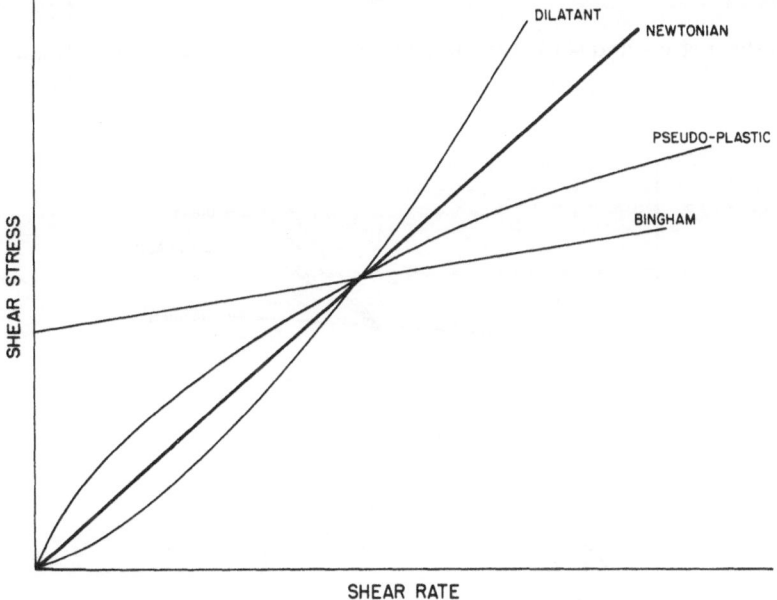

Fig. 12.   Flow curves of Newtonian and non-Newtonian systems.

the tongue, action of swallowing, etc; therefore, what is probably felt in the mouth is some type of continuum, with either shear or time as one axis, and viscosity or stress as the other axis. It has been shown that the sensory feel of 'sliminess' could be related to the rate at which the viscosity of a system decreases with an increasing rate of shear (Szczesniak and Farkas, 1963).

In many cases, the flow behavior of a non-Newtonian system follows the power equation:

$$\tau = K\dot{\gamma}^n + C$$

where τ = shear stress (dyne cm^{-2}); $\dot{\gamma}$ = shear rate (s^{-1}); K = consistency index; n = flow behavior index; and C = yield stress.

This equation is very useful and quite widely used to characterize liquid and semi-liquid food products. The constant 'n' denotes the degree of deviation from Newtonian flow, i.e. the degree of curvature of stress/shear curves. Since Newtonian products exhibit no yield ($C=0$) and stress which is directly proportional to strain ($n=1$), the above equation reduces to:

$$\tau = \eta\dot{\gamma}$$

where $\eta=$ coefficient of viscosity.

For Bingham products which show Newtonian flow once the yield point has been reached, the equation becomes:

$$\tau = \eta\dot{\gamma} + C.$$

When shear stress and shear rate data are plotted on log log coordinates, a straight line relationship is obtained as shown in Figure 13. The slope of the line gives the

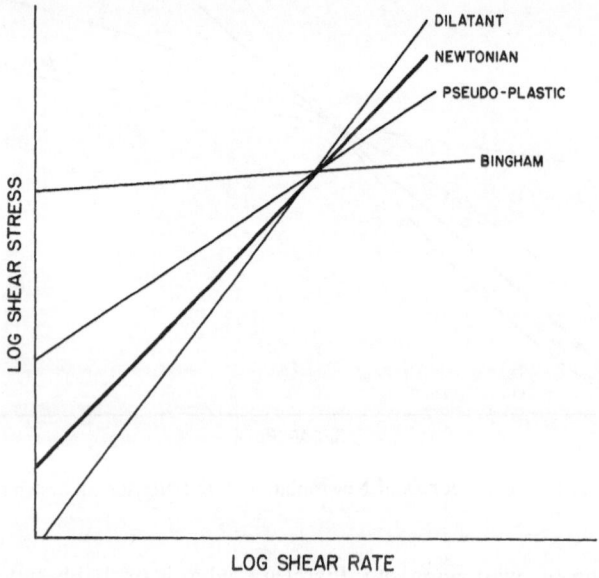

Fig. 13.   Double logarithmic plots of flow curves.

value of 'n' (1 in the case of Newtonian flow) and the intercept with shear stress axis quantifies the consistency index for pseudoplastic ($n>1$) and dilatant ($n<1$) materials with no yield point. For products that exhibit a yield, the intercept reflects both K and C. Determination of fundamental consistency constants with various types of viscometers has been discussed by Charm (1960, 1962, 1963).

Similarly to the situation with solid foods, the trend in viscometric characterization of liquid foods is towards more sophisticated, better designed, recording instruments.

The Haake Rotovisko viscometer is finding more widespread applications (e.g. Charm, 1963) and a recording coaxial cylinder viscometer which draws a continuous shear rate/shear stress flow curve has been recently described in the literature (Corey and Creswick, 1970).

Probably the most popular capillary viscometer used in testing liquid food products is the Ostwald Viscometer modified by Cannon and Fenske. It is a glass capillary viscometer operated by gravity and operable only in low shear-stress ranges. Basically, it consists of reservoir bulbs and a capillary U-tube arrangement. The measurement involves the determination of time needed for a fixed volume of liquid to flow a given distance under an exactly reproducible hydrostatic head. Some of the sources of error in this type of a device are: wall effects, surface tension, improper drainage of the bulb, elastic stresses, and viscous end effects.

A brief mention should also be made of in-line viscometers used for quality and process control. These devices are designed for continuous viscosity measurements. They are installed directly in process vessels or lines eliminating the need for sampling. Most respond to both Newtonian and non-Newtonian products. Typical of that class of instruments is the Dynatrol viscosity control device. It consists of a U-shaped metal probe which vibrates at 120 cycles per second and with an amplitude dependent on the viscosity of the liquid flowing around it. When viscosity increases, the amplitude decreases due to an increased shear resistance of the medium. The output signal is converted into a 0–10 MVDC signal compatible with recorders or controllers.

3.8. EXTRUSION DEVICES

In general, extrusion measurements have not been as popular as some of the other tests discussed; however, they have been used as an index of textural quality for viscous liquids, gels, fruits, vegetables, and fats. A thorough review of extrusion devices and methodology used for measuring textural parameters of various foods has been made by Bourne and Moyer (1968). These researchers concluded that:

A wide range of sizes and shapes of test cells have been used for extrusion testing. The maximum force, average force, total work done, weight of material extruded, and extrusion time have all been used as the measured variable.

An imitative-type test based on the principle of extrusion is the butter spreader (Prentice, 1954). It is a screw extruder in which a cube of butter is extruded and sheared by a knife edge at standard loads and speeds. The amount of shear required is said to indicate spreadability, but correlation with organoleptic scores has been only good to fair.

Fortini and Hogan (1966) extruded cooked potatoes through an ordinary kitchen ricer and reported that the required force may be a good measure of mealiness. Extrusion methods have also been used to quantify consistency of fats (e.g. Pyhala, 1935; Scherr and Witnauer, 1967). A test for meat tenderness based on essentially the same principle has been considered (Sperring et al., 1959). Lanza and Kramer

(1967) used the extrusion method for assessing graininess of apple sauce and Angel *et al.* (1965) used it on peas.

Several different types of cylindrical extrusion cells recently have been constructed by Kramer and Hawbecker (1966) for use with the Kramer Shear Press. They employ either a tight-fitting round plunger, or one which is 1/4″ smaller in diameter than the cell. In the latter case, a 1/8″ annulus is created through which 'back extrusion' can take place when the cell is fitted with a solid bottom. Grid-like bottoms or bottoms with an orifice are employed with the tight-fitting plunger (see Figure 6b).

Bourne and Moyer (1968) published an in-depth study of back extrusion using fresh green peas as the test material. The study showed that extrusion results in a constant force which is not affected by sample size nor the presence of added water. The annulus width is important in that it has an inverse effect on the recorded force, affects the resolution between samples, and the influence of plunger speed on force readings. A 4 mm annulus was reported to give the best resolution. Because of its sensitivity, absence of friction, ease of construction, maintenance and cleaning, the method appears to offer promise in routine testing.

3.9. MISCELLANEOUS

The ingenuity and the imagination of researchers have created many other instruments which often represent intriguing approaches to mechanical food testing. Only a few examples will be cited.

The Alveograph has been used for studying the textural properties of dough by measuring the work done in forming and bursting a bubble of dough.

The succulometer measures juiciness of the food as the volume of the liquid expressed from a known weight of material at a given pressure in a given time (Kramer and Smith, 1946). The results have been used to judge maturity of corn.

The food mincer has been used to measure toughness of meat as the work required to grind a given amount of material to a given particle size (Miyada and Tappel, 1956).

A rather unique empirical test was devised recently (Anon., 1968) for measuring textural properties of Swiss rolls. The sample is held in a sleeve across which can pass a hook. The force required to 'bite' through the roll with a hook when the sleeve is pulled downwards is said to correlate well with sensory acceptability tests.

Tensile strength measurements have been applied to measuring toughness of meat and other products composed of distinct fibers. They have also been used to test fruit skin (e.g. tomato) and toughness of strings in string beans. The main problem in tensile testing is how to grip the specimen in such a way as not to weaken it at the point of contact with the clamps while at the same time assuring a firm grip. Pool (1967) has published on a method which overcomes that problem. He used a special adhesive (Eastman 910, methyl-2-cyano acrylate) to bond the two ends of a cylindrical poultry meat specimen to flat metal plates which were then connected to the tensile strength testing apparatus. Force (or work) required to tear the sample was related to the tenacity of connective tissue. Cooked meat samples were cut uniformly in such a manner that the fibers were parallel to the plane ends of the cylindrical specimen.

Another problem in tensile strength measurements of solids is the uneven distribution of stresses and chance failure. Cutting the specimen in the form of a dumbbell can overcome these errors.

A method for determining the tensile strength of fluid food materials was published by Charm (1964). Products such as ketchup, mayonnaise, tomato paste, etc., are forced slowly from a vertical tube 0.4 cm in diameter. The ratio of the weight of the column to the cross-sectional area at the break point is taken as the tensile strength.

Recently, attention has been drawn to the application of dynamic tests. Their advantages appear to be: short response time, low levels of stress, and non-destructiveness. Correlation with sensory evaluation remains to be established and may become the most serious drawback. Pioneering work in this area was done by Drake (1962), Finney and Norris (1968), and Abbott et al. (1968a, b). Their techniques are based on using sonic energy to vibrate the specimen; internal friction, Young's modulus and resonance frequency (taken as a measure of stiffness or firmness) are calculated from the specimen's response. Because of its non-destructive character, the test can be performed repeatedly on the same sample, therefore, permitting one to follow changes on maturing (e.g. on the tree), storage, aging, etc., without introducing errors due to sample-to-sample differences. Correlation of these tests with sensory evaluation will hinge on our better understanding of how dynamic elastic properties relate to sensory qualities. A recent attempt to correlate the results from the resonance technique to eating qualities of apple fruit resulted in the conclusion that the pressure test (destructive in nature) was a more reliable index of sensory firmness (Finney, 1971).

3.10. MULTIPLE-PURPOSE UNITS

As the name implies, these are instruments which can be used for performing a number of different texture tests. They have gained considerable popularity in recent years because of their versatility, flexibility, accuracy, and appealing design features.

Most popular in the group are the Instron Universal Testing Machine and the Food Technology Corporation's Texture-Test System (formerly the Kramer Shear Press). Less sophisticated stress-strain testing units similar to the Instron are being manufactured by several firms including W. C. Dillon, Thwing-Albert, Chatillon, etc.

The current popularity of this type of equipment follows the thought expressed in Section 2.1 that all the devices used for measuring textural parameters have a number of elements in common. This thought was further developed by Voisey (1971) who focused attention on the test cell (i.e. the probe in contact with the tested foodstuff) as the key element. All the other elements can be built into a single universal unit which can serve a number of different purposes when fitted with different test cells.

Originally designed for measuring physical properties of metals, rubbers, fabrics, packaging materials, etc., the Instron Universal Testing Machine (Figure 14) is

Fig. 14. The Instron Universal Testing Machine with compression box
(courtesy of Prof. M. C. Bourne).

composed of a mechanical drive system, a load cell to measure forces generated either
in compression or tension, a recorder, and a set of controls to automate the performance of the unit and introduce considerable flexibility and versatility. The mechanical
drive system consists of a horizontal crossbar (called the crosshead) driven vertically
up and down by two lead screws. Tensile tests are performed above and compression
tests below the moving crosshead under most experimental conditions. The crosshead

can be driven at a number of speeds. The recorder chart is driven by a synchronous motor through a set of gears. These can be the same or different from the crosshead gears, thus allowing the expansion or contraction of the distance axis. The advantages of this unit are: rigid construction, freedom from friction or deformation in the basic machine, constant speed, precise knowledge of the values of both force and distance (or time) axis on the recorder chart, high precision and accuracy, and results given in fundamental calibrated units of force, distance, and time (or their functions). In addition, the instrument can be preset for maximum distance or maximum load at which the movement of the crosshead stops; it can be programmed for automatic return, cycling, relaxation tests, etc.

There is virtually no limit to the type of test cells that can be employed with the Instron. Some are supplied by the manufacturer; others must be obtained from other sources or constructed by the researcher himself. Crushing, shearing, puncturing, extrusion, flexure, compression, etc., tests can be easily performed.

The Texture-Test System (Figure 6a) is composed of a Texture Press into which are mounted the test cells, a hydraulic power system, and a force measuring system consisting of a proving ring and a force readout dial or a transducer. In the latter case, a recorder is also used to provide a force-distance tracing. The recorder chart drive is connected through a mechanical linkage to the hydraulic cylinder power ram, thereby resulting in a constant ratio of the two speeds. The various types of test cells available with the Texture-Test System are shown in Figure 6b. Extrusion, compression, back extrusion, single-blade shear, and multiple-blade shear tests can be performed with these cells. The speed of ram travel (rate of force application) can be varied by adjusting the control valve which governs the flow of the hydraulic fluid. A reservoir provides for heat dissipation to maintain uniform fluid temperature. Whether this is really accomplished in practice has been questioned by many researchers who claim some changes in ram speed resulting from the increasing temperature of the driving fluid during continuous testing. The Texture-Test System does not provide for as great a versatility or accuracy as the Instron but, nevertheless, is a useful, practical tool which finds much application in quality control work.

4. Standardization of Instruments

The problem of standardization and calibration has not yet been solved (see Twigg, 1960, for a discussion of the problem). The use of more sophisticated instruments, recorders, and control devices has allowed standardization of parts of the apparatus concerned with the mechanical movement and force detection. Recommended procedures have been discussed by Voisey (1971); however, there are at present no satisfactory means for standardizing the test cells other than comparison with a reference cell.

Materials such as asbestos, spun protein fibers, beeswax, weatherstripping, filter paper, etc., have been proposed and used with some degree of success by various researchers; however, this success has usually been limited to the laboratory advocating the method, and none of these materials have found a general acceptance. The procedure practiced with the Pea Tenderometer, to standardize the entire apparatus including the test cell using a standard lot of the test food, appears to have much practical appeal. To this day, Pea Tenderometers across the country are being calibrated early in the harvest season using sub-samples from the same lot of peas. It is obvious that such procedures are cumbersome, not very "scientific", and leave room for innovations.

It appears that much could be gained in this area by combining the efforts of food technologists interested in objective texture measurements with those of high polymer chemists. It is conceivable that high polymer materials could be identified that would exhibit mechanical properties of the right intensity and type. A suitable standardizing material should be cheap, easily available, reproducible from lot to lot, and free from changes due to aging, exposure to ultraviolet light, small moisture losses, and small temperature variations.

Recently, Stanley (1970) identified a microcrystalline, soft wax (Ultraflex Amber) as a potentially suitable standardizing agent for the Pea Tenderometer. Following the development of a refined technique for preparing uniform slabs, Voisey and Nonnecke (1971) used this material for testing a number of Tenderometers at processors' plants in Ontario and confirmed its promise. Repeatable results were obtained provided instrument readings were corrected for temperature effects which influence the firmness of the wax.

As stated by Kramer (1969), "any test procedure, whether it be subjective or objective, should have a high degree of precision, accuracy, and calibratability." Since in most cases objective methods are desired that will relate to sensory evaluation, it is generally agreed that instrumental tests do not have to be more accurate than sensory tests. What is badly needed, however, are significant advances in psychophysics which will provide a better understanding of the sensory perception and quantification of textural parameters.

5. Current Trends

Several significant trends are quite apparent to those who, in the last decade or so, have been watching closely the activities in the area of instrumental texture measurements.

First, there has been an obvious shift from commodity-oriented to principle-oriented type of research. No longer are instruments being designed for specific products, but rather elucidated principles are being applied to measurements of specific products. No longer is the interest in pea texture limited to horticulturists and that in meat texture to biochemists. With an understanding and appreciation of common problems, a sense of communality has developed which contributes to the sharing of advances and speeds progress towards needed solutions.

Secondly, texture research is beginning to be treated as a subject in itself, rather than as a by-product of quality assessment, product development or breeders programs. With several research groups following this approach in an interdisciplinary manner, significant strides have been made towards a better understanding of the concept of texture and of more adequate methods of instrumental texture measurements.

Old methodology is being subjected to critical scrutiny and empirical approaches are gradually being modified (or substituted) with more theoretically sound principles. The availability of better testing equipment coupled with the application of physical principles had already allowed to subject several empirical test methods to a critical analysis.

Another significant trend is to center attention on the test cell, i.e. the probe in contact with the food, and use the same driving, force-detecting, and force-recording mechanism with a number of different cells. This trend was first brought into focus with the application of the Instron Universal Testing Machine to food texture work. Even prior to this, the Kramer Shear Press was equipped with several different cells in addition to the standard multi-blade, shear-compression cell; therefore the concept of 'universal units' has been born as an approach to food texture measurements.

'One-point' measurements are giving way to 'multiple-point' measurements, i.e. to recordings of the force/distance (or stress/strain) relationships from which not only force values at a given distance point can be determined but changes in force (if any) over a range of distances can be observed.

With the realization that texture is a spectrum of different parameters, the principle of 'texture profiling' has been developed and is gaining popular acceptance. In this approach, force-distance curves recorded during the test are being analyzed in terms of several parameters correlatable with sensory evaluation. The measurement is performed in such a way that a meaningful separation into several component characteristics can be made. Measurements which involve a composite of different forces do not easily lend themselves to this type of an analysis.

The principle of 'texture profiling' has recently been extended to include a combination of measurements obtained from several different instruments or from several different tests with the same instrument. Using factor analysis, these instrumental measurements can then be correlated with a spectrum of sensory parameters (Henry et al., 1971; Toda et al., 1971). This approach is very useful not only for obtaining better correlations between instrumental and sensory measurements but, more importantly, for developing a better understanding of the meaning of sensory texture assessment in terms of physical correlates.

Last, but not least, is the trend towards a sounder basis for correlating instrumental with sensory evaluation. Four separate lines of endeavor are being felt here, not totally successful yet, but nevertheless, worth noting: a general study of stress/strain relationships during the process of mastication, equating instrumental testing conditions for specific products with those occurring during sensory testing, translating sensory reactions into basic rheological parameters, and performing objective texture measurements under strictly defined conditions, thereby permitting calculations of basic rheological parameters.

Appendix 1. Named Instruments Described in the Literature

Name	General type	Reference
Adams Consistometer*	radial flow	Adams and Birdsall (1946)
Armour Consistometer	rotating tube viscometer	Harrington (1948)
Armour Tenderometer	penetration	Hansen (1972)
Asco Firmness Meter	squeezing	Garrett et al. (1960)
ASTM Grease Penetrometer*	penetration	Rich (1942)
Baker Compressimeter*	compression	Am. Assoc. Cereal Chemists, 1962, 'Cereal Laboratory Methods', method 74-10
Ball Compressor*	compression	Caffyn and Baron (1947)
Bloom Consistometer	torque	Bloom (1938)
Bloom Gelometer*	penetration	Bloom (1925)
Babb Compressimeter	compression	Babb (1965)
Bostwick Consistometer*	linear flow	Davis et al. (1954)
Boucher Jelly Tester	compression	Koprowski (1951)
Brinell Hardness Tester*	compression	Lovegren et al. (1958)
Butter Extruder*	extrusion	Prentice (1954)
Butter Consistometer	spreading and cutting	Kapsalis et al. (1960)
Carbide Penetrometer	biting	Simon et al. (1965)
Child-Satorius Shear	shearing	Satorius and Child (1938)
Christel Texturemeter*	multiple-probe puncture	Christel (1938)
Chopin Alveograph*	bubble pressure	Chopin (1936)
Chopin Teleplastometer	kneading	Chopin (1965)
Crusell Consistometer	extrusion	Pyhala (1935)
Curd Firmness Tester	cutting	Voisey and Emmons (1966)
Dassow's Shear-Jaw Device	shear	Dassow et al. (1962)
Dexometer	biting	Lehmann (1907)
Dough Extensimeter	extension	Haltor and Scott Blair (1937)
Fiberometer	cutting	Kramer et al. (1949)
Firm-O-Meter	squeezing	Kattan (1957)
Gel Characterization Apparatus	compression	McAllister and Reichenwallner (1959)
General Foods Texturometer*	biting	Friedman et al. (1963)
Geneva Texturometer	puncture	Moyer et al. (1956)
Grünewald Tenderometer	biting	Grünewald (1957)
Hamden Deformation Tester	compression	Rowlands (1964)
Hoeppler Consistometer	compression, penetration	Hoeppler (1954)
Hutchinson Cone Penetrometer	penetrometer	Haighton (1959)
Karlsruhe Device	twist	Nemitz (1963)
Kramer Shear Press*	compression/shear (presently, multiple cells)	Kramer (1961)
'KT' Biting Device	biting	Kelly et al. (1960)
Lundstedt Curd Tester	penetration	Lundstedt (1955)
MacMichael Viscometer	rotational shear	MacMichael (1915)
Magness-Taylor Pressure Tester*	puncture	Magness and Taylor (1925)

Appendix 1 (Continued)

Name	General type	Reference
Martin Consistometer	rotational shear	Martin (1941)
Martin Tenderometer	shear	see Pea Tenderometer
Maturometer	puncture	Mitchell *et al.* (1961)
McCollum Firmness Meter	puncture	Garrett *et al.* (1960)
Minnesota Shear Stress App.		see Child-Satorius Shear
MIT Denture Tenderometer	biting	Brody (1957)
Monsanto Resonance Elastometer	oscillation	Slater (1954)
New Jersey Pressure Tester (forerunner of the Magness-Taylor Pressure Tester)	puncture	Clark (1928)
Panimeter	compression	Hintzer (1949)
Pea Tenderometer*	shear	Martin (1937)
Penn State Testing Machine	compression	Mohsenin (1963)
Plastometer	in-line tube viscometer for pseudoplastic materials	Eolkin (1957)
Platt Compressimeter (forerunner of the Baker Compressimeter)	compression	Platt (1930)
Plumit	penetration	Rich (1942)
Recording Dynamometer	compression	Babb (1965)
Ridgelimiter*	sag	Lockwood and Hayes (1931)
Rigidimeter	displacement	Scott Blair and Burnett (1957)
Rigorometer	extensibility	Briskey *et al.* (1962)
Rotating Knife Tenderometer	cutting	Bouschart and Meyer (1965)
Sectilometer		see Curd Firmness Tester
Shortometer	compression	Davis (1921)
Slice Tenderness Evaluator (STE)	puncture	Kulwich *et al.* (1963)
Smith Tensiometer	tensile strength	Smith (1957)
Succulometer*	expulsion of fluid	Kramer and Smith (1946)
Swift's Tenderness Testing Device	compression	Palmer (1962)
Tarr-Baker Delaware Jelly Tester	compression	Tarr (1926)
Torsiometer	torsion	Southorn (1960)
Variable Pressure Viscometer	extrusion	Scherr and Witnauer (1967)
Volodkevich Bite Tenderometer	biting	Volodkevich (1938)
Warner-Bratzler Shear*	shearing	Bratzler (1932)
Winkler Device	biting	Winkler (1939)
Wolodkewitsch Apparatus	different cells	Wolodkewitsch (1956)

* denotes commercially available devices

Appendix 2. Commercially Available Instruments

Instrument	Manufacturer or Distributor[a]	Approx. price[b] (U.S. $)
Adams Consistometer	Central Scientific Co. 2600 St. Kostner Ave. Chicago, Ill. 60623	120
Alveograph	M. Chopin and Cie 5, Rue Escudier Boulogne-sur-Seine France	1400
Baker Compressimeter	Watkins Corp. P.O. Box 445 Caldwell, N.J.	550
Biscuit Texture Meter	Boher Perkins (Exports) Limited Westwood House 13 Stanhope Gate Park Lane, London W1, England	700
Bloom Gelometer	Precision Scientific Co. 3739 W. Cortland St. Chicago, Ill. 60647	400
Bostwick Consistometer	Central Scientific Co. 2600 S. Kostner Ave. Chicago, Ill. 60623	100
Brabender Amylograph	C. W. Brabender Instruments Inc. 50 E. Wesley St. So. Hackensack, N.J. 07606	2300–2700
Brabender Consistometer	C.W. Brabender Instruments Inc. 50 E. Wesley St. So. Hackensack, N.J. 07606	950
Brinell Hardness Tester	Testing Machines, Inc. 398 Bayview Ave. Amityville, N.Y. 11701	260–890
Brookfield Viscometers	Brookfield Engineering Labs., 240 Cushing Street Stoughton, Mass. 02072	400–700
Burrell Extrusion Rheometer	Burrell Corporation 2223 Fifth Ave. Pittsburgh, Pa. 15219	360
Butter Consistometer	The Accurate Manufacturing Co. 945 King Ave. Columbus, Ohio	
Butter Tester	N.V. Apparatenfabriek van Doorn Utrechtseweg 364 De Bilt, Holland	230–1980
Chatillon Testers	John Chatillon & Sons Div. of Aero-Chatillon Corp. 83–30 Kew Gardens Road Kew Gardens, N.Y. 11415	200–500
Cherry-Burell Curd Tension Meter	Manton Gaulin Mfg. Co. 67 Garden Street Everett, Mass. 02149	250

Appendix 2 (Continued)

Instrument	Manufacturer or Distributor[a]	Approx. price[b] (U.S. $)
Christel Texturometer	Seifert Manufacturing Co., Inc. Kiel, Wisc. 53042	2000
Do-Corder	C.W. Brabander Instruments Inc. 50 E. Wesley St. So. Hackensack, N.J. 07606	7200
Durometer	Shore Instrument Co. 90–35 Van Wyck Expressway Jamaica, N.Y.	70–100
Dynatrol Viscometer	Automation Products, Inc. 3030 Max Roy Street Houston, Tex. 77008	2500–3500
Egg Deformation Tester	Marius Ganzenmarkt 4–8 Utrecht, Holland	400
Extensigraph	C. W. Brabender Instruments Inc. 50 E. Wesley St. So. Hackensack, N.J. 07606	4400
Farinograph	C. W. Brabender Instruments Inc. 50 E. Wesley St. So. Hackensack, N.J. 07606	3000
FIRA Jelly Tester	H. A. Gaydon & Co., Ltd. 93 Lansdowne Road Croydon, England	445
FIRA/NIRD Extruder	H. A. Gaydon & Co., Ltd. 93 Lansdowne Road Croydon, England	765
Fruit Pressure Tester	Effe-Gi Corso Garibaldi 102 48011 Alfonsine Ravenna, Italy	30
Gel Tester	Marine Colloids, Inc. P.O. Box 70 Springfield, N.J. 07081	450
General Foods Texturometer	Zenken Co., Ltd. Kyodo Building No. 5, 2-Chome Honcho Nihonbashi, Chuo-ku Tokyo 103, Japan	7800
Haake Rotovisko Viscometer	Brinkmann Instruments, Inc. Cantiague Road Westbury, N.Y. 11590	4000
Hill Curd Tester	Heusser Instrument Mfg. Co. 123 W. Malvern Ave. Salt Lake City, Utah 84115	350
I.B.V.T. Hardness Meter	N.V. Ledoux Daluragen 47 Dodewaard, Holland	1620
Instron Universal Testing Machine	Instron Corp. 2500 Washington St. Canton, Mass. 02021	4–20000

Appendix 2 (Continued)

Instrument	Manufacturer or Distributor[a]	Approx. price[b] (U.S. $)
Kramer Shear Press	Food Technology Corp. 12300 Parklawn Drive Rockville, Md. 20852	see Texture Test System
MacMichael Viscometer	Fisher Scientific Co. 711 Forbes Avenue Pittsburgh, Pa. 15219	625
Magness Taylor Pressure Tester	D. Ballauf Co. 619 H. Street N.W. Washington, D.C. 20001	45

"Mechanical Thumb" –
   attachment for the above
	Agricultural Specialty Co.
	   (ASCO)
	11313 Frederick Ave.
	Beltsville, Md. 20705
		65

Mitex Bending Tester	Imass, Inc. P.O. Box 134 Accord, Mass. 02018	4000
Nametre Acoustic Spectro- meter	Nametre Co. 272 Loring Ave. Edison, N.J. 08817	6000–10600
NIRD Pitching Machine	Packman Research, Ltd. Twyford, Berks. England	115
NIRD – Plint Cheese Curd Torsiometer	Plint & Partners, Ltd. Fishponds Road Workingham RG11 20G Berks., England distributed in the U.S.A. by CAA Scientific P.O. Box 1234 Darien, Conn.	1700
Ottawa Food Texture Meter	Canners Machinery, Ltd. Simcoe, Ontario, Canada	1740–4050
Pabst Texture Tester	R. E. Pabst Co. 5115 Westheimer Houston, Tex. 77027	4–5400
Pea Tenderometer	Food Machinery & Chemical Corp. Canning Machinery Division 103 E. Maple St. Hoopeston, Ill. 60942	2900

Penetrometers	(a) The Hutchinson Instrument
	       Company Ltd.
	       Suffolk House, George Street
	       Croydon, Surrey, England
		250

	(b) Precision Scientific Co.
	       3739 W. Cortland Street
	       Chicago, Ill. 60647
		250–620

	(c) Lab-Line Instruments, Inc.
	       Labline Plaza
	       Melrose Park, Ill. 60160
		380–680

Appendix 2 (Continued)

Instrument	Manufacturer or Distributor[a]	Approx. price[b] (U.S. $)
	(d) VEB Feinmess Dresden 8023 Dresden Kleiststrasse 10 Deutsche Demokratische Republik	
Plasti-corder	C. W. Brabender Instruments, Inc. 50 E. Wesley Street So. Hackensack, N.J. 07606	7700
Plint Cheese Curd Torsiometer	Plint & Partners, Ltd. Fishponds Road Workingham RG11 20G Berks., England	see NIRD – Plint Cheese Curd Torsiometer
Ridgelimeter	Sunkist Growers, Inc. Product Sales Div. Ontario, Calif. 91764	90
SCA Tablet Hardness Tester	Strong Cobb Arner, Inc. subs. of International Chemical & Nuclear Corp. 11700 Shaker Blvd. Cleveland, Ohio 44120	280–600
Succulometer	Food Technology Corp. 12300 Parklawn Drive Rockville, Md. 20852	sold only as an attachment to the Texture Test System
SUR Penetrometer	Sommer and Runge KG Bennigsenstr. 23 Berlin 41 German Federal Republic	120–280
Tensile Tester/Tensilon	Tokyo Measuring Instruments Co., Ltd. No. 104, 1-Chome Chotumine- machi Ota-ku Tokyo, Japan distributed in the U.S.A. by Imass, Inc. P.O. Box 134 Accord, Mass. 02018	4750
Texture Test System	Food Technology Corp. 12300 Parklawn Drive Rockville, Md. 20852	5300–7500
Torry Brown Homogenizer	H. G. Brown Electronics Pty., Ltd. Lower Mills Bugby – Clarkston Glasgow, Scotland	800
TVC Cream Corn Meter (similar to Adams Consistometer)	The United Company TVC Road Westminster, Md. 21157	150
UC Fruit Firmness Tester	Western Industrial Supply, Inc. 236 Clara Street San Francisco, Calif.	265

Appendix 2 (Continued)

Instrument	Manufacturer or Distributor[a]	Approx. price[b] (U.S. $)
Visco-corder	C. W. Brabender Instruments, Inc. 50 E. Wesley Street So. Hackensack, N.J. 07606	1100–1600
Viscoelastometer/Rheovibron	Tokyo Measuring Instruments Co., Ltd. No. 104 1-Chome Chotumine- machi Ota-ku Tokyo, Japan distributed in the U.S.A. by Imass, Inc. P.O. Box 134 Accord, Mass. 02018	5700–15 500
Warner-Bratzler Shear	G-R Electric Mfg. Co. Route #2 Manhattan, Kan. 66502	400
Weissenberg Rheogoniometer	Sangamo Weston Controls, Ltd. Bognor Regis, England distributed in the U.S.A. by The Martin Sweets Co. 3131 W. Market Louisville, Ky 40212	17–23 000 19–25 000

[a] inquiries should be directed to respective manufacturers or distributors.
[b] these represent information obtained at the time this list was compiled and are for guidance only.

CHAPTER VII

INDIRECT METHODS OF OBJECTIVE
TEXTURE MEASUREMENTS

ALINA SURMACKA SZCZESNIAK

1. Introduction

A number of indirect methods have been employed by various researchers for objective texture measurements. These include chemical, enzymatic, microscopic, and various physical techniques other than those involving the use of some form of mechanical energy. The latter were covered in the preceding chapter.

All of these indirect methods must assume that there is a definite relationship between texture and the tested characteristic, and that there are no modifying effects exerted by other parameters. Since this is seldom the case, this type of approach to texture measurements has definite restrictions and has not gained widespread popularity, except in a few cases. Meat and plant tissues have been the focal test materials.

2. Chemical

The use of chemical methods for estimating texture of vegetable crops has been described in some detail by Kramer and Twigg (1970), while applications to meat texture investigations have been reviewed by Szczesniak and Torgeson (1965).

Moisture content and alcohol insoluble solids are the two most widespread chemical methods used as textural indices for fruits and vegetables. They are actually measures of maturity, but – since texture changes as the plant material matures – they became accepted as texture indicators.

2.1. MOISTURE CONTENT

In 1939, Caldwell showed that moisture decreases and solids increase as vegetables mature and become tougher and less desirable. Moisture determination has become a standard quality control and maturity index for many vegetables, especially sweet corn. The official AOAC method (7.003) for determining the percent moisture in plant tissue involves drying at 95–100°C in a vacuum oven. Because this is time consuming (about 5 h) and often not very reliable, other methods have been used such as toluene distillation, moisture balances, conductivity measurements, etc. As has been stated by Joslyn (1970), "the accurate determination of moisture, in some respects the most simple of analytical operations, is frequently one of the most difficult determinations which the food chemist is called upon to make". This is because of errors introduced by possible decomposition of the product and loss of

A. Kramer and A. S. Szczesniak (eds.), Texture Measurements of Foods, 109–117. All Rights Reserved.
Copyright © 1973 by D. Reidel Publishing Company, Dordrecht-Holland.

volatile food constituents which affect the weight loss. Distillation with toluene eliminates the latter but not the former error. It is also subject to inaccuracies arising from an incomplete recovery of water due to emulsion formation or clinging to the sides of the apparatus (ibid). The official AOAC method (7.004) involves combining the test sample with toluene in a distillation apparatus, distilling and collecting the immiscible water/toluene mixture, and determining volumetrically the amount of condensed water.

2.2. ALCOHOL INSOLUBLE SOLIDS (AIS)

This method, introduced in 1935 by Kertesz, involves extracting the sample with 70–80% ethanol and determining the weight of the washed, dried residue (AOAC method 32.006). The residue is composed of starches, celluloses, fiber, pectins, and proteins. It increases in amount as the texture of the plant material tends to toughen on aging. The method is particularly suitable for peas, sweet corn, and lima beans that, at maturity, contain substantial quantities of soluble solids and for which a measure of total solids is not a valid index of maturity (or tenderness). The Food and Drug Administration has adopted it as an official method in the standards of quality for canned peas and sweet corn.

The method is so widely used for these two vegetables that almost every published article on textural aspects of peas and sweet corn contains data on AIS, and every attempt at developing a new instrumental method of texture evaluation for these products judges the quality of the technique on the basis of its degree of correlation with AIS values. Working with peas, Bourne and Moyer (1968) demonstrated a good correlation with the extrusion method, Gutschmidt (1954) with the Christel Texture-meter, and Casimir et al. (1971) showed a relationship with the puncture technique. Gould et al (1951) published extensive data on the relationship between AIS and various parameters of texture and growing conditions for yellow sweet corn.

Many instances, however, have been demonstrated of serious limitations of AIS as an index of textural quality. The method falls short when significant alterations in firmness occur as a result of enzymatic activity (e.g. bruising of cherries, effect of vining juice on peas), or of cations (mainly Ca) added during processing. In these cases, de-esterification of pectin and formation of a calcium pectinate gel network lead to considerable increases in firmness without significant effects on AIS. In a very informative discussion of the chemical basis of texture, Isherwood (1955) pointed out that factors affecting texture (cell wall structure and soluble and non-soluble metabolites) are all interrelated and textural changes are not caused by any one factor alone.

It may be stated in summary that AIS may be a good comparative index of maturity of plant tissue, but – unless all other factors are equal – is not always a reliable indicator of textural quality.

2.3. FIBROUSNESS

Many vegetables, such as asparagus, wax and green beans, develop a significant fiber content as maturation progresses. This lowers their eating quality and makes them less desirable from the consumers' point of view. The fibers contribute to an overall toughness of the product and often can be felt distinctly in the mouth during mastication, especially if the surrounding tissue is tender. The role of fibrousness in orally perceived texture was discussed by Matz (1962).

The fiber content can be determined by measuring the residue undigested by boiling in acid or alkali. Kramer (1951) developed a method based on maceration of a sample in a Waring Blender, filtering, drying, and weighing. This method is quite rapid and is claimed to be more directly related to fibrousness as sensed organoleptically. The description of a washing apparatus for use in fiber determination of green beans was published recently by Kaldy *et al.* (1967).

2.4. CRUDE FIBER

This determination assesses the total cellulose and lignin content in plant tissue disregarding the geometry important in fibrousness. The official AOAC method (7.053) defines crude fiber as "loss on ignition of dried residue remaining after digestion of sample with 1.25% H_2SO_4 and 1.25% NaOH solutions under specific conditions." The crude fiber content increases as vegetables lose their tender, succulent character and become overmature.

Similarly to AIS, the fiber content is only an indirect method of texture evaluation and in many cases has been shown to be unrelated to the overall product toughness (e.g. Dunlop and Ormrod, 1970).

2.5. CONNECTIVE TISSUE

Chemical determination of connective tissue in meat has received much attention because of its claimed close relationship to tenderness. Fractionation with dilute NaOH, estimation as alkali-insoluble proteins and the hydroxyproline determination have been the main methods used (Szczesniak and Torgeson, 1965). The last technique is considered the most accurate. It is based on the fact that collagen (which occurs in connective but not in muscle tissue) is uniquely high in hydroxyproline and may be quantified by means of a color reaction.

There is much disagreement concerning whether chemical determinations of connective tissue are a valid indicator of meat tenderness. From an extensive review of the subject, Szczesniak and Torgeson (1965) concluded:

that the research results reported so far do not offer a clear picture regarding the usefulness of chemical methods for connective tissue determination as a measure of meat tenderness. This may be due partly to the limitation of the methods themselves and partly to the more complicated relationship between connective tissue and tenderness. As Deatherage had put it: 'there is little evidence to indicate that classical proximate analyses of chemical entities can give us a direct measurement of tenderness ... Meat is a biological system and gets much of its character through the organization of various constituents.

More recently, attention has been shifted to the role of muscle (myofibrillar) proteins and sarcomere in meat tenderness and the chemical determination of connective tissue has fallen essentially into disuse.

Other chemical methods that have been used include pectin, starch, alkaline earth metals, pH and extractable nitrogen determinations. Of these, pectin analyses, as applied to plant tissue, have received the most attention. Textural changes in apples, peaches, and other fruits during maturation and storage have been related to increases in water-soluble pectins and decreases in insoluble fractions (e.g. El-Sayed et al., 1966; Shewfelt, 1965; Woodmansee et al., 1959). This alters the mechanical resistance of cell walls and the adherence of the cells leading to softening and a general decrease in resistance to applied forces. Pectic substances have also been shown to be involved in determining viscosity characteristics of tomato products and fruit juices. Analytical determinations of this type are usually restricted to research work. They are too complicated, and the relationship to sensory texture too obscure to merit more widespread use.

3. Enzymatic

Enzymatic methods of texture evaluation have been applied mainly to meat. Similar to some of the other methods already discussed, their primary use has been as research techniques rather than as routine evaluation methods. Various digestive enzymes such as papain, bromelain, ficin, trypsin and Rhozyme P-11 were employed. Adams et al. (1960) compared the enzyme digestion with instrumental (Warner-Bratzler Shear), sensory and chemical (collagen determination) methods of beef tenderness evaluation. They obtained significant correlations for cooked meat and no apparent relationships for raw meat.

Enzymatic methods require extensive training and experience on the part of the operator. Additional criticisms are based mostly on the lack of enzyme specificity (muscle proteins vs. connective tissue) and on inadequate knowledge of the biochemistry of meat tenderness.

4. Microscopic

4.1. NON-STAINING TECHNIQUES

Direct counting of ruptured starch cells in dehydrated potato granules has been used to predict gluiness and undesirable mouthfeel qualities of the reconstituted product. Hall and Fryer (1953) preferred this to a number of other methods evaluated, including viscometric procedures.

In many other foodstuffs (e.g. tomato products, Whittenberger and Nutting, 1957), the microscopic appearance of the material has been shown to relate to consistency as sensed organoleptically or as quantified by mechanical means. Because microscopic examination is time-consuming, requires an experienced observer, and is a very indirect method of texture characterization, it has found application mostly in research work.

A microscopic technique which is currently receiving much attention is the determination of the length and diameter of sarcomere in muscle, factors claimed to be involved in meat tenderness (e.g. Herring et al., 1965).

4.2. HISTOLOGICAL METHODS AND STAINING TECHNIQUES

Histological methods involve expert tissue preparation and, thus, suffer from similar but more intense disadvantages than the direct microscopic evaluation of unfixed samples mentioned above. Their very important advantage over chemical methods, however, is that they give information not only on the presence or absence of certain tissue components known to affect texture but also on the geometrical arrangement and structure. An excellent review on the relationship of histological structure to the texture of fresh and processed fruits and vegetables has been published recently by Reeve (1970) and the reader is referred to that paper for an informative discussion of the subject. Reeve stated that "because texture concerns the manner in which the parts of a substance comprise the whole, the origin of textural qualities... is essentially histological."

Histological and histochemical techniques have been applied to studying textural aspects of meat and plant tissue. They involve killing, fixing, and solidifying the tissue so that it can be cut into very thin sections. Tissue slices are then stained, mounted on slides and studied microscopically. Light, phase, and electron microscopes have been employed. Paul (1949) published an excellent review of the fundamentals of histological methods used with meats, while the basic techniques used with plant tissues can be found in the textbook by Johansen (1940).

5. Physical

In this section will be considered miscellaneous physical methods (most of which are specific for certain foodstuffs) and non-destructive instrumental methods not previously described in the chapter on 'Instrumental Methods of Texture Measurements'.

5.1. DENSITY

Brine separation is a popular practical method of separating for processing fresh shelled peas and other seed vegetables into density groups believed to correspond to texture groups. A similar method is used for potato tubers. This method's disadvantage, besides its own errors, is that it is only a quick, indirect method of determining total solids that are not always a textural or maturity index. Nevertheless, in the case of fresh green peas, the method has practical use, although its validity has been questioned (e.g. Boggs et al., 1943). In the case of some other plant foodstuffs, e.g. watermelon (Showalter, 1961), the specific gravity method has no recognized value for determining maturity.

La Belle (1964) published concerning bulk density measurements and concluded that values for precooked dried beans, blanched apple slices, and sweet and red tart cherries were related to texture differences arising from harvesting, handling, or processing variables. The advantage of the method is that it can measure a large quantity of material non-destructively. Besides its indirectness, a disadvantage is the error inherent in the subjective judgment of when the desired level of fill has actually been reached.

Bulk density and apparent true density (taking into account the pore volume) were used, among other tests, to describe textural characteristics of compressed foods (Morris, 1965).

In the area of baked goods, specific gravity of batters has been associated with good volume and textural quality of the finished product (Funk *et al.*, 1969) and recommended as an indirect quality control tool. Specific gravity of batters is determined as the ratio of the weight of a given volume of batter to the weight of an equal volume of water.

Since fat has been claimed to be a factor in meat tenderness, various methods for determing fat in meat have been proposed. One of these is based on an instrumental analysis of specific gravity of deboned meat (Whitehead, 1970).

5.2. VOLUME

This physical parameter has been shown to relate to texture of bread, cakes, and other baked products, other factors being equal. The most popular method of determining the volume of baked goods is by seed displacement. In this method, the volume of the test sample is obtained as the difference between the volume of rape seeds in an empty container and that in the same (or equivalent) container holding the sample. Funk *et al.* (1969) discussed the advantages and disadvantages of this method and described some of the other suggested procedures.

5.3. CELL STRUCTURE AND OPENNESS

The volume of baked goods is an indirect method of determining the openness of the cell structure that is the predominant factor in textural quality. More direct methods for describing the cell structure have been proposed. Swartz (1938) described a method involving the determination of the weight of sand held by the cell structure following rotation on a 40° incline as a quantitative measure of cell size or openness of grain. The more open the structure, the more sand is retained. The validity of this method and irregularities arising from nonuniformity of texture in different parts of the cake were confirmed by Funk *et al.* (1965). Absorption of kerosene, ink prints, and shadowgrams have also been used to quantify openness of the structure.

5.4. DRAINED WEIGHT

Drained weight of canned or frozen/thawed products is sometimes used as a quick index of textural damage caused by processing (e.g. Leonard *et al.*, 1958; La Belle and Moyer, 1960; Bradley, 1966). The measurement reflects fluid retention. Injured tissue becomes soft and flaccid and exudes much liquid leading to low values for drained weight. In performing the measurement, the product is allowed to drain on an eight mesh sieve for two minutes and its weight is then determined. The weight of the sample prior to processing must be known. The drained weight measurement may be quite misleading when the fruit is packed in syrup because of sugar retention in the flesh on one hand, and the osmotic effect on the other hand. Board *et al.* (1966) reported that drained weights of canned strawberries and loganberries decreased as the syrup concentration increased. Addition of calcium salts and of colloids (e.g. carrageenin, low and high methoxyl pectins) increases the drained weight by firming the texture in the former case and binding the exudate in the latter case (e.g. Wegener *et al.*, 1951; Barton, 1953).

5.5. DRIP

This measurement is similar to that of drained weight, but it involves the determination of the volume of exuded liquid instead of the weight of the solid phase. It is used mostly with frozen products, especially fish. Factors affecting the formation of drip in fishery products have been reviewed by Miyauchi (1963).

Drip formation on thawing is associated with quality deterioration because (a) drip by itself is unsightly and undesirable, (b) it reduces the weight of the product, and (c) it is the manifestation of tissue damage leading to textural changes. As discussed by Milleville and Leinen (1962), conventional freezing of fish destroys the firm, gelatinous texture and causes the formation of a spongy, fibrous texture. This fibrous texture cannot retain the fluid which exudes, leading to drip. Treatment of fish with polyphosphates prior to freezing significantly reduces the thawing drip (e.g. Mahon and Schneider, 1964); however, factors minimizing drip do not necessarily result in a desirable texture (e.g. Buttkus and Tarr, 1962).

5.6. PRESS FLUID

While drip measurements involve the determination of fluid volume exuded from the test sample under gravity, those of press fluid are based on application of mechanical force.

Press fluid reflects the water-binding capacity of meat and bears a positive relationship to the sensory parameter of juiciness. Various techniques for determining the press fluid of meat were reviewed by Szczesniak and Torgeson (1965). The most widely used is the method of Grau and Hamm (1953). It consists of placing the meat sample on filter paper contained between two plates, applying pressure, and measuring the area wetted by the expressed liquid. Szczesniak *et al.* (1963b) adopted this method for use with the General Foods Texturometer.

With vegetables such as sweet corn, the amount of juice expressed by a given force is taken as an index of maturity and succulence. The more juice expressed, the younger and more desirable is the corn. A convenient method for measuring the press fluid of vegetables is the use of the succulometer cell developed for the Kramer Shear Press (presently known as the Texture Testing System). In this method, the sample is placed in a round test cell with a spout, force is applied mechanically, and the amount of expressed fluid is collected and measured in a graduate cylinder (Kramer and Smith, 1946).

5.7. MISCELLANEOUS

A variety of other physical methods have been proposed for quantifying various textural parameters in an indirect manner.

Textural changes of fruit upon maturation have been assessed by means of light transmittance techniques (Birth and Norris, 1958; Ernest *et al.*, 1958; Romani *et al.*, 1962) and electrical resistance measurements have been applied to butter as a rough measure of texture (Prentice, 1953). Use of X-ray diffraction techniques was proposed to trace the progress of staling in bread and concomitant increases in firmness.

Eggshell fragility has been measured non-destructively by means of atomic radiation. Known as 'beta backscatter', the technique involves firing beta particles at the egg and counting the bounce-back. A tough shell returns more particles than a fragile one (James and Retzer, 1967).

A U.S. patent has been issued recently for a device that, when attached to an individual fruit, gives a distinct color change when the fruit reaches maturity and an edible state. The device consists of a thin layer of a CO_2 absorbent and a pH color indicator sandwiched between two pieces of Mylar. On ripening, CO_2 is given off and produces a color change in the pH indicator (Davis, 1969).

Sag of unmolded gels was measured by obtaining a series of shadowgraphs on a photographic plate. Graphic translations of the shadow lines gave objective values useful in the study of melting and relaxation characteristics of gelatin, pectin, alginate, and other gel types (Pintauro and Lang, 1959).

Since meat texture has received more attention than that of any other foodstuff, the literature offers many references to various indirect physical methods of meat texture evaluation (Szczesniak and Torgeson, 1965). Color of meat, pH, diameter of muscle fibers, various carcass measurements, and even hair density were considered, most without much success.

A turbidimetric method developed by Love (1960) appears to have an application to measuring toughness of fish. It is related to the observation that toughness in fish is connected with decreased fragility of muscle cells. A sample of the muscle is disintegrated mechanically and the optical density of the solution is determined at 430 mμ. High optical density indicates tender fish that yield cells easy to disintegrate. More recently, Partmann (1971) proposed a turbidimetric method for quantifying protein denaturation of fibrillar proteins (involved in muscular contraction) as a means of detecting textural changes during frozen storage of white muscle of cod, carp, and chicken. A suspension of fiber fragments in a suitable salt medium is reacted with adenosine triphosphate and turbidity is then measured optically using a green filter. The method has not yet been related to either sensory or mechanical texture measurements.

A novel method based on sounds produced during sensory chewing of food has been worked out by Drake (e.g. 1965a and 1965b). Although of potentially high value to fundamental research on food texture and its sensory perception, the method is not yet applicable to more routine quantification of textural parameters.

6. Concluding Remarks

It is clear from the foregoing discussion that, with few exceptions, indirect methods of texture evaluation have been limited to research work and have not found widespread practical applications. The few exceptions are exemplified by AIS and moisture content of vegetables. It is agreed, however, that these are criteria of maturity and not necessarily those of texture.

Chemical and histological methods are very valuable tools in obtaining a better understanding of fundamentals of texture. As has been pointed out in Chapter V, they have contributed much to our present day knowledge of the basic elements of texture and of factors causing desirable or undesirable changes and hopefully will continue to do so in the future. Much can also be gained from combinations of these methods with instrumental and sensory techniques to derive a better understanding of what is involved in texture perception.

CHAPTER VIII

PSYCHOPHYSICAL AND PSYCHOMETRIC
MEASURES OF TEXTURE

HOWARD R. MOSKOWITZ, BIRGER K. DRAKE and CAJ ÅKESSON

1. Introduction

Sensory evaluation of texture in foods, as well as in other products such as textiles, belongs to the domain of psychology known as *psychophysics*. Psychophysics directly concerns the correlation of sensory experience with physical measures. Its ultimate goal is to establish mathematical equations or relations that permit the scientist to predict the sensory characteristics of materials from the physical measurements, and vice versa. Psychophysics does not seek to explain the basis of sensation, but instead searches for orderly relations between the subjective realm and the physical world. Much has been written about the development of machines to quantify the physical characteristics of food that give rise to 'texture' as perceived by the human judge. However, machines cannot appreciate texture like a human being. Psychophysics of texture thus has two goals: a description of the relevant physical characteristics of materials that are implicated in the perception of texture, and an analysis of subjective-objective correlations to relate one or more physical characteristics to the ultimate percept experienced by the human judge.

The present review provides a foundation for the psychophysics of texture by indicating and discussing the available sensory scales that may be used by the human observer. Traditionally, psychophysics has viewed the human observer mechanistically, as a machine that transduces physical information supplied by the environment into sensory and perceptual information experienced by the subject. Sensory scales in this context provide the scientist with a transfer function, which allows him to determine how quantitative relations between physical measures are translated into relations between subjective percepts.

Prior to introducing the types of scaling procedures used by psychophysicists it is relevant to briefly discuss the theory of measurement. Two major problems have been considered by philosophers and psychologists interested in quantifying sensory experience (cf., Suppes and Zinnes, 1963): the problem of *representation*, which can be stated as the justification of assignment of numbers (or other signs) to stimulus objects in order to represent magnitudes of sensory properties; and the problem of *uniqueness*, which refers to the degree to which this assignment is unique.

A solution of the representational problem of measurement consists of a complete characterization of the empirical operations used in the measuring procedure and of proving that these operations are isomorphic (or possibly homomorphic) to ap-

A. Kramer and A. S. Szczesniak (eds.), Texture Measurements of Foods, 118–129. All Rights Reserved.
Copyright © 1973 by D. Reidel Publishing Company, Dordrecht-Holland.

propriately chosen numerical operations. It should be pointed out that an adequate solution of the representation problem is obtained not merely by demonstrating the existence of a numerical relational system isomorphic to a given empirical relational system. In addition, the numerical system must conform to the real number system. To illustrate the problem of representation, let us consider the instrumental measurement of 'extensive' properties like length, mass and time. A formal measurement theory for such properties is a set of assumptions or axioms formulated by using an ordering operation R and a 'concatenation' operation $\oplus$. These two operations permit the construction of a scale m such that for any two objects a and b: (i) aRb if and only if $m(a) \geq m(b)$; and (ii) $m(a \oplus b) = m(a) + m(b)$. For the properties of length, mass and time, the empirical interpretation of 'R' and '$\oplus$' is obvious. It is possible to prove (e.g., Suppes, 1951; Suppes and Zinnes, 1963) that any empirical 'extensive' relational system, i.e., a relational system consisting of a set of objects together with an R and a $\oplus$ operation, is isomorphic to a numerical relational system consisting of the set of positive numbers together with the arithmetical operations '$\geq$' and '$+$'. Hence, the representational properties of measurement of length, mass and time are easily demonstrated. For many psychometric procedures, on the other hand, there are no empirical operations available for describing the relational structure of the stimulus set. In such cases, there can be no explicit formulation of any set of axioms specifying the necessary and sufficient conditions for a numerical representation of the stimuli, and the representational problem remains unsolved. Unfortunately, a direct proof of the existence of ordering and concatenation operations valid for, e.g., a texture perception, would require psychometrical procedures rather different from those available. However, indirect conclusions can be reached by examining, with current psychometrical methods, the structure of response sets.

A solution of the uniqueness problem consists of determining the scale type that can be derived from the measurement data. Following Stevens (1951) one can distinguish between four basic types of scales: nominal, ordinal, interval and ratio scales.

The weakest type of scale is the nominal scale, where numbers are used simply to name the stimuli. The assignment of numbers is then completely arbitrary and does not convey any information on magnitudes of properties.

For the ordinal scale, the assignment of scale values is arbitrary except for order. Every monotone transformation of the scale values is admissible. The ordinal scale with natural origin has the additional restriction that the number '0' is assigned to zero amount of the property to be measured. Within the field of physics, Moh's hardness scale for minerals and the Beaufort wind scale are well-known examples of ordinal scales.

For interval scales, only linear transformations are admissible. Neither the unit of measurement nor the zero point can be determined in a non-arbitrary way. The measurement of temperature in degrees Fahrenheit or Celsius (centigrades) is an obvious example of an interval scale.

For ratio scales, the assignment of scale values is non-arbitrary except for the choice of unit of measurement. This is equivalent to the statement that a ratio scale is unique up to multiplication by a positive constant. In physics, this type of scale may be exemplified by the measurement of 'extensive' properties like length, mass and time.

In addition to these 4 scale types, which are sufficient for a wide range of practical applications, many others are conceivable. For instance, difference scales which are unique up to an additive constant, and hyperordinal scales (Coombs, 1952) for which hypermonotone transformations preserving first differences are admissible, have been suggested.

Psychophysical scales must always be subjected to scrutiny with these two problems in mind. Quite often the scientist may lose perspective and fail to model the most appropriate physical relations on his scale. He may also lose sight of permissible transformations which will not change the quantitative, formal properties of the scales.

2. Dimensions of Perceived Texture

Empirically, both subjective and objective (instrumental) assessments of texture are multidimensional in nature. No single instrument reading can fully characterize the rheological properties of a food material, and even untrained observers will report a complex of sensations that occurs when they chew and swallow a food. Keywords such as 'sliminess', 'firmness' and 'unctuousness' may be sufficient to elicit quite different responses from the observer as he samples a pudding. As attention focuses on different subjective attributes of texture, the importance of physical characteristics shifts so that some become more relevant and others diminish in importance.

An appropriate first step in the development of psychophysics of texture is to classify by suitable methods the linguistic descriptions that are most relevant, and then to specify how these terms may be used to denote unambiguously the salient characteristics of texture. Two general approaches have been made in this direction to extract the underlying 'dimensions' important to the observer. The first is psychological in nature and seeks to extract fundamental dimensions of texture perception without necessarily referring to instrumental measures. The second begins with an assumed set of basic dimensions believed to characterize subjective perception of texture and then seeks analogues in physical measures.

Within the field of psychometrics a number of so-called multidimensional scaling techniques have been developed for analysing sets of stimuli which, simultaneously, vary along different dimensions. Such methods are generally based on similarity judgements, which can be obtained by instructing the subjects to assign a number to the perceived degree of similarity for each stimulus pair, or by asking them which of two stimulus pairs has the higher degree of similarity. Alternatively, one may obtain indirect measures of similarity by examining the degree to which different stimuli are confused. In any case, the analysis of similarity data is based on geometrical models, in which the stimuli are considered as points in some multidimensional metric space (in most cases a Euclidean space). The judgements of similarity are assumed to reflect distances between the stimuli in that space.

The classical multidimensional scaling methods, such as factor analysis and related techniques (Harman, 1967), require numerical similarity judgements as input data. The similarity judgements are conceived as ratio-scaled distances in a Euclidean space of unknown dimensionality. By varying the number of dimensions of the metric space one tries to find the configuration of stimulus points whose distances most closely correspond to the numerical input values. The theoretical basis of factor analysis is a theorem proved by Young and Householder (1938) stating the necessary and sufficient conditions for embedding a set of points in a Euclidean space.

In the methods of ordinal multidimensional scaling the judgements are assumed only to reflect the order of interstimulus distances in some multidimensional metric space. The idea of such ordinal methods is to recover the configuration of distances and coordinates of the stimuli using only ordinal-scaled interstimulus distances. Ordinal multidimensional procedures were first developed by Coombs (1952), and a variety of computational methods have been devised for analysing orders of similarity judgements (e.g., Kruskal, 1964; Lingoes, 1966; Torgerson, 1965).

Yoshikawa et al. (1970) presented a factor analysis of Japanese descriptive terms for texture, and showed how these terms applied to a selected group of named foods. Factor analysis (Harman, 1967) condenses the different descriptive terms to basic dimensions presumed to be independent of each other. The large array of words uncovered by Yoshikawa et al. appeared to be composed of different weightings of the following basic underlying dimensions: hard-soft, cold-warm, oily-juicy, elastic-flaky, heavy, viscous and smooth. Each descriptive term for texture could be described by a linear combination of the dimensions, but the fundamental dimensions were assumed to be pure. No attempt was made as yet to correlate these dimensions to instrument measures.

Multidimensional analysis provides a useful technique for uncovering the existence of simple geometric representations of texture perception. The goal is to represent the attributes of subjective texture as a series of axes in multidimensional space (a space of seven dimensions for the Japanese analysis). A given texture term may then be expressed as a linear combination of these fundamental dimensions. More powerfully, however, the texture of a food may be represented by assigning that food to a point in the multidimensional space whose coordinates provide the relative weights of each of the seven dimensions. A consideration of multidimensional representation ought to be made for instrumental measures in terms of fundamental rheological principles, much in the same way as was done for subjective attributes.

Szczesniak and her colleagues (Szczesniak et al., 1963a) have provided an illustration of the basic psychophysical approach to classifying texture attributes. The system is based upon an appreciation of instrumental measures of texture, as well as ancillary dimensions that may participate in and influence the perception of texture. Each dimension is selected in order that it may be related to an instrumental measure, and assessments may be effected either by a panel of observers or by an appropriately constructed machine.

Scott Blair (1966) has also considered the construction of texture scales based upon instrumental measures. In one series of studies in which observers were instructed to judge 'firmness', he noted that their judgments could best be predicted by a combination of physical dimensions in a manner not permitted by physical principles. These 'intermediate entities' as he called them, suggest that a psychophysical approach to characterizing the dimensions of texture ought to consider the possibility that observers may combine rheological dimensions in a manner not consonant with standard physical principles.

An intensified study of psychophysical approaches deserves as much interest as measurement techniques, for it may provide to the instrumentalist a systematic and relatively straightforward way of reducing texture to analyzable components in terms of subjective percepts. The future of these approaches, however, depends critically upon the continued refinement of psychological methods to extract the nuances of texture perception.

3. Scales of Magnitude in Texture

For the past 150 years psychophysicists have struggled with the problem of quantifying experience, and the subjective attributes of texture are no exception. A list of relevant attributes or dimensions is only part of the task of describing sensations. Two food materials may possess the same set of characteristics, but yet may seem entirely different because the characteristics are present to different degrees in each food. A short history of the attempts of psychophysicists to construct intensity scales is relevant for an appreciation of those commonly used today.

In the early part of the nineteenth century, E. H. Weber (Boring, 1942) reported the results of his studies on the discriminability of stimuli. In a study of the resolving power of the touch sense, Weber discovered that two stimuli were reported to differ in sensory magnitude only when the more intense one exceeded the weaker one by a relatively small, but fixed, proportion of physical intensity. For example, two weights appeared to differ in heaviness if one was approximately 10% heavier. If they differed by a smaller ratio, they were reported as being equally heavy. In order to represent this critical fraction, Weber chose the expression $\Delta I/I$ (later called the 'Weber Fraction') to signify the smallest fractional part by which a stimulus had to be increased in order to render the two stimuli just noticeably different.

Subsequently, in the middle of the nineteenth century, a German physicist G. T. Fechner (Boring, 1942) suggested that a sensory scale could be constructed by adding units of discriminability in a differential equation. The result was a logarithmic function $S = k \log I$ that could be used to relate the sensory magnitude S to the physical intensity I. The logarithmic relation was an attractive conjecture since it set forth a simple and tractable mathematical transformation to relate increments of physical magnitude to those of perceived magnitude.

A complete method based on Fechner's assumptions is tedious since it involves adding together many units of subjective discriminability in order to span the range of sensation from 'barely perceptible' to 'extremely strong'. Fortunately, more direct and simpler methods have since become available, and their use has gained for psychophysics a new appreciation of the utility of intensity measurements.

Psychophysicists in the past fifty years have suggested that sensory measurement be approached in a direct fashion by requesting the observer to report his perceived magnitude for a given stimulus along an appropriate response scale. In this way, the judgments of an observer are treated as would be the output of a measuring instrument, and the technique of arriving at data is simple and immediate. Because the observer carries with him the experience of a lifetime in making informal judgments of magnitude, several caveats must be followed. Often the order of stimuli in a sensory test can influence the outcome of the measurement, as the observer recalls the judgments he had made for previous stimuli. Many observers exhibit peculiar number behavior and consistently refuse to use the extreme portions of the scale, even to describe intense or faint stimuli (Stevens and Greenbaum, 1966). Perhaps the most important caveats to follow are that the observer be aware of the dimension to which he is instructed to attend and to judge, and that he report his judgment in a manner appropriate for the response scale that has been selected.

Three major methods have been used to arrive at measures of magnitude: ordinal scaling, interval scaling, and ratio scaling. The convention method of ordinal scaling may be dismissed immediately, because it provides only an indication of the relative position of stimuli along an intensity continuum. No information is forthcoming about the relative separation of different stimuli and, therefore, no unique equations or functional relations can be derived. Interval and ratio scales can provide this information.

4. Interval Scales of Texture Perception

Scales of perceived texture may be developed by partitioning an appropriate sensory continuum into a series of ordered categories (e.g., 1–9, 1–100), and then instructing the observer to assign to each physical stimulus a category rating by selecting the category that best describes the perceived magnitude. For example, a 1–9 scale of perceived 'hardness' may be of interest. At the low end of the scale (category 1), the anchor word 'very soft' may be used, whereas at the high end of the scale (category 9) the anchor word 'very hard' would be appropriate. For each sample to be judged, the observer selects the most appropriate category. On occasion, intermediate verbal anchors may be inserted at one or another category to aid in making judgments.

Interval scales rely upon the concept of sensory 'distances'. In order to use the scale properly and to apply statistical tests to the subjective responses, the differences between adjacent categories in the scale must be equal. Thus, for example, the subjective hardness difference between categories 3 and 4 must be equal to the hardness difference between categories 4 and 5. If this criterion is met, the scale values for different textural attributes may be analyzed by the methods of inferential statistics to determine whether or not two materials differ significantly in hardness (e.g., by the t test, or for more samples by the technique of Analysis-of-Variance). It is appropriate to conclude that two samples are separated by 4 units on the hardness scale when the category values for these samples are 6 and 2, respectively. Because the zero point is not fixed, however, one cannot conclude that the harder sample exceeds the softer one by a factor of 3. Thus, differences between category (interval) scale ratings convey information about intervals, but proportions or ratios cannot be obtained.

The 'Texture Profile' (Szczesniak *et al.*, 1963a) comprises a series of category scales for the subjective dimensions of hardness, brittleness, chewiness, gumminess, adhesiveness and viscosity. Each attribute is represented by a scale containing a series of ordered categories, although the number of categories is not the same for each dimension. There are nine categories of hardness, eight of viscosity, seven each of brittleness and chewiness, and five each of gumminess and adhesiveness. Each scale is anchored by a series of representative standard products that illustrate the intensity gradations for the subjective magnitude on that dimension.

The value of category scales such as the 'Texture Profile' lies in the possibility of describing functional relations between perceived magnitude along the subjective scale and instrumental measures. Szczesniak *et al.* (1963a) present six graphical functions that relate the perceived intensity of texture to instrumental readings of the General Foods Texturometer. In all cases but one (subjective viscosity), the category scale arrived at in the profile could be described either by a simple linear function $S = kT + b$ (S = subjective rating, T = Texturometer reading), or a logarithmic function $S = k \log T + b$.

5. Ratio Scales of Sensory Texture

Many physical scales of magnitude possess the convenient property that ratios of measures convey meaningful information about proportionality. For example, 2 miles represent twice the distance of 1 mile, and 40 grams represent five times the weight of 8 grams. Category scales lack this ratio property, and although category scales can be very sophisticated in their usage, they nonetheless must be considered inadequate when proportionality is of interest.

During the 1940's, Beebe-Center and his colleagues (Beebe-Center and Waddell, 1948) developed four scales for the primary tastes of salty, sour, sweet and bitter. They used a technique entitled 'fractionation', which was as follows. A top concentration of a representative taste compound (e.g., sodium chloride for 'salty') was selected, and the observer was instructed to sample that aqueous concentration, and then to select from a series of comparison samples the solution tasting half as salty. Successive applications of this procedure produced an array of concentrations of NaCl that appeared subjectively to correspond to the numbers 100:50:25:12.5 etc., (equally 2.5:1.25 etc.). A later report (Beebe-Center, 1949) listed nine standard concentrations of chemicals for each of the four primary tastes. The 'gust scale', as Beebe-Center called it, set the stage for numerous subsequent investigations of other sensory and perceptual continua.

In a study of subjective roughness and smoothness, Stevens and Harris (1962) instructed their observers to assign numbers in proportion to the perceived intensity by the technique of *magnitude estimation*. In essence, the observers acted as meters to generate numerical output, with the assumed property that ratios of numerical assignments to stimuli corresponded to subjective ratios of perceived magnitude. Subsequent studies investigated the sensory functions for many other senses (Stevens, 1960). Other experiments directly relevant to the psychophysics of texture are those by Ekman *et al.* (1964), also on the perception of roughness and smoothness as well as on the acceptability of those sensations.

Harper and Stevens (1964) studied the relation between physical hardness of rubber (measured by the degree of indentation of a weight on the rubber sample) and perceived hardness. They reported that a power function with an exponent of 0.8 related the subjective magnitude to the objectively measured degree of hardness. Subjective perception of viscosity was studied by Stevens and Guirao (1964) who instructed their observers to either stir a viscous silicone oil while blindfolded, stir it while looking, or simply turn the solution over in an enclosed container. Moskowitz (1972) also investigated subjective viscosity and fluidity, but with twelve vegetable gums made up in solutions of different concentrations.

The techniques that provide ratio scales are simple to use and give reproducible results across different laboratories. Over a period of about two decades, beginning in 1953 (Stevens, 1953), there has been a continued and increasing interest in appropriate equations to relate sensory and objective (instrumental) measures. Consistently, equations of the form $S=kI^n$, the simple power function, appeared adequate to describe the results of these direct scaling experiments, whereas the simple linear, logarithmic or exponential functions appeared to be unsatisfactory.

A major problem that arises in direct scaling of any type, and especially in ratio scaling where functional relations are of extreme importance, is the validity of the scale. How can one assert that the texture scales arrived at by magnitude estimation truly obey the properties of a ratio scale wherein proportionality statements are meaningful? In a validation study, Ekman *et al.* (1960) demonstrated that the technique of magnitude estimation provides essentially the same scale as is obtained when the observer is asked to judge directly the ratio of intensity between every pair of stimuli in the set (monochromatic wavelengths in that case). Although individuals may possess idiosyncratic ways of handling numbers as measuring instruments, and there appear to be systematic biases resulting from the selection of the initial stimulus to serve as a standard and the initial judgment to serve as a modulus (scale unit), the outcome of these direct experiments is a robust scale that can be replicated in different laboratories. It should be noticed, however, that the calculation of ratio scale values from magnitude estimation data often requires complicated computational procedures (cf. Poulton, 1968; Sjöberg, 1966, 1971), and there is no such procedure available that is generally valid.

The present main use of power functions is to index how rapidly sensory magnitude grows with increases in physical intensity. For example, values of n equaling 1.0 indicate a linear relation between sensory and objective measurements, so that doubling the physical intensity in turn doubles the subjective impression of magnitude. Line length viewed on a wall (Stevens and Guirao, 1963) is such a continuum. On the other hand, when n is less than 1.0, as is the case for subjective loudness with $n = 0.64$ (Stevens, 1966), or subjective viscosity with $n = 0.4$ (Moskowitz, 1972) or $n = 0.5$ (Stevens and Guirao, 1964), the subjective magnitude grows more slowly than the physical intensity, and the range of subjective intensity is narrower than that measured physically. An example is provided by the viscosity exponent of 0.5. If two gums are dispersed in water and their viscosities are 1000 and 100 cps respectively, then their apparent viscosities lie in a ratio of 10:1. The observer, however, in judging the sub-jective viscosity of these two gums will tend to rate them as a ratio of only $(10/1)^{0.5} = 3:1$ (approximately). Finally, n may exceed 1.0, in which case the opposite effect occurs, so that the subjective magnitude grows more rapidly than the physical intensity and the subjective range of magnitude exceeds the physical range. The smoothness of sandpaper is governed by a power function with an exponent of 1.5, when grit size is used as the instrumental measure (Stevens and Harris, 1962). Thus, when the grit of sandpaper is increased tenfold, the sensory effect is a 'jump' of apparent smoothness by a factor of $(10/1)^{1.5} = 30:1$ (approximately).

The power function of sensory magnitude provides a simple index of perceived magnitude for different aspects of texture. The exponent n is independent of the units of instrumental and subjective measurements, so that instrument readings that vary by a multiplicative factor (i.e., change of unit) will still yield the same power function exponent when correlated with subjective judgments. In addition, n also remains un-affected when the observer multiplies all of his judgments by a constant factor, al-though on occasion the size of numbers that an observer selects will tend to bias his judgments. Some observers feel comfortable using small numbers, and may not make an equivalent multiplication for all of their judgments when changing their scale to large numbers. The coefficient k is, however, extremely sensitive to changes in the size of numbers used by the observer.

6. Applications of Sensory Scales of Magnitude

One aim of sensory scaling is to replace the instrumental measure with the human observer, or *vice versa*, to replace the human observer with one or more instrumental measures. Two of the most important areas that can benefit from developments in quantification of subjective perception are the fields of quality control and of product development.

Quality control can delimit the physical characteristics of products so that only acceptable and safe foodstuffs are offered for consumption. No one wishes to eat rock-hard hamburgers or mushy apples, and our decision to accept or to reject these items is based upon the measuring capacity of our senses. Knowledge of psychophysical relations, whether based upon interval (category) or ratio scale methods, may provide improvements in quality control.

Continued research by Kramer (1965, 1969) represents a fertile avenue of applica-tion. An illustrative study is provided by the experiment of Fox and Kramer (1966), who attempted to predict the overall quality and flavor of cooked and raw green beans. Their instrumental measures were the shear press, the qualitometer and other 'objective' indices, while the sensory evaluations were provided by both trained and untrained panelists. Their interest was to determine an equation that could predict with high accuracy the sensory judgment (S) as a function of the instrumental measures. The technique of 'multiple regression' (Efroymsen, 1960) was used to arrive at the following linear relation between sensory magnitude (S) and instrumental readings $X_1, X_2...X_n$

$$S = k_1 X_1 + k_2 X_2 + k_3 X_3 \cdots + k_n X_n.$$

The independent, *ad hoc* predictor variables, $X_1, X_2,... X_n$, were weighted multipliers $k_1...k_n$, respectively, in order to maximize the predictability of the sensory response (S).

Kramer's equations are designed to provide a statistically satisfactory prediction of subjective responses to texture, but are not focused upon any theoretical model of human responses to texture variables. A valuable contribution to the psychophysics of texture perception would be a combination of Kramer's technique with standard psychophysical scaling, such as the Texture Profile. That combination would bring together the analytical techniques with statistical prediction theory, and would provide a firm basis for the study and employment of sensory and instrumental texture measures.

Another challenging possibility is that acceptability limits in quality control may be expressible both in terms of instrumental measures and in units of sensory magnitudes. For example, under appropriate standardization of scaling techniques, one can assign to a tomato a firmness rating in proportion to a reproducible standard of firmness, as well as measure the 'instrumental firmness' by objective techniques. If the standard sensory firmness is called 10, and the tomato feels half as firm, then the tomato is rated as 5. Furthermore, if – on this sensory scale – acceptably firm tomatoes are rated between 4 and 9, with other ratings indicating tomatoes that are too hard or too mushy, then it is possible to set up equivalent limits of physical measurement and to define the region of acceptability in terms of instrumental measures. For example, if subjective firmness is governed by a power function of the form $F = I^{2.0}$ (i.e., subjective firmness F grows as the square of instrumental reading I), then the upper level of subjective acceptance (9) corresponds to an instrument reading of 3, and the lower level (4) to an instrument reading of 2. By this technique, it is possible to employ both instruments and human observers in the quality control process, and to make them interchangeable according to a transformation function between sensory and instrumental magnitude.

Product development work is another area that can use psychophysics of texture. Developing a product requires the close cooperation of food technologists who do the actual development work with psychologists who evaluate the consumer reaction to the product. The Texture Profile may be of great use in providing a system to characterize the salient texture dimensions of a food. For the past fifteen years the Hedonic Scale of food acceptability (Peryam and Pilgrim, 1957) has provided quantitative information which permits the development team to assess the prospective acceptability of their product. Neither procedure, however, possesses the capacity to assess both the dimensions of a food that need to be changed in order to improve acceptability, and the magnitude of change that should be effected. A combination of the Texture Profile, to focus attention on salient dimensions, with direct questions on the desired magnitude of change would provide the much needed approach towards more effective changes in the texture of newly developed foods.

A recent study (Moskowitz, 1971) has focused on this problem of 'optimizing' foods by instructing the taste panel to judge both the intensity of a sensory dimension (the subjective 'grind' of hamburger) and the degree to which the panel would like to alter that dimension (i.e., increase or decrease the subjective grind) in order to optimize the acceptability. The initial phase of the experiment concerned the relation between perceived 'chunkiness' (the subjective attribute) and the objective measure of grind. The relation could be described by a power function of the form $C = kG^{0.55}$ ($C =$ subjective chunkiness, $G =$ physical measure of grind in inches). The panelists were then instructed to estimate the degree to which they would increase or decrease the grind of each hamburger in order to make the overall product maximally acceptable. Since a psychophysical equation was determined, changes along the subjective dimension could be related to changes in the physical degree of grind, and – therefore – the textural property of the hamburger could be altered in accordance with subjective estimates.

The work of Lundgren (1969) in assessing the 'hand' of textiles is appropriate in the application of psychophysics to the study of texture. According to Lundgren, textile hand is an integrated property of several sensory responses such as roughness, stiffness, bulk and thermal properties. In order to assess the appropriateness of these dimensions for a product, Lundgren proposes a multi-stage procedure, wherein the initial steps are to instrumentally measure magnitudes of each dimension, and to determine the response profile of the human tester. A combination of these two measures, one objective and the other subjective, is an index of how the tester responds to the specific textile characteristics. The outcome is a final profile of the product in terms of the acceptability of each dimension to the tester, with the profile serving as a quantitative index of which aspects must be changed in order to optimize the final acceptance of the fabric.

7. Outlook

Psychophysical scaling and sensory evaluation are in their infancy. The opportunities they present for providing scientific and technological information on ways to assess the dimensions of sensory texture are limited only by the ingenuity of the scientist and of the product developer. The pioneering work to assess intensity on many sensory continua by the method of ratio scaling (Stevens, 1960) as widely different as the loudness of noise, the hardness of rubber and the sweetness of sugar, promises that corresponding interest in food will also reveal important theoretical and practical functions in quantifying sensory responses.

One may, thus, look forward to the coming decades with confidence, as holding the promise of answers to many of the important questions in evaluating the subjective attributes of texture. Developing textures in accordance with consumer specifications, producing better descriptions of the salient dimensions of products, and finally arriving at precise sensory-instrumental correlations are all research areas that will no doubt profit greatly from the advancement of texture psychophysics.

CHAPTER IX

QUANTIFICATION OF OBJECTIVE AND SENSORY TEXTURE RELATIONS

JOHN G. KAPSALIS, AMIHUD KRAMER, and ALINA S. SZCZESNIAK

1. Introduction

Relating instrumental with sensory methods of texture measurement is of vital interest for several reasons. The first and the most important is the need to develop objective tests which could predict, and ultimately substitute, for the tedious and time-consuming sensory evaluation. Such tests would have the obvious advantages of greater speed, ease of standardization, freedom from drift, and other complicating factors associated with panel testing. Since texture is a sensory attribute, the second reason is the desire on the part of the researchers to explain instrumental measurements in terms of human sensations, i.e., relating what the apparatus detects to what is being felt during handling or consumption of the food.

In correlating sensory with objective measurements, it is imperative that both types of tests be performed with utmost precision. Since texture is a sensory attribute, the sensory, i.e., panel method is assumed to be accurate while the objective, i.e., the instrumental method may be accurate only to the extent that it agrees with panel results. Lack of precision, or bias, in panel data, however, can and does occur, so that too often well-designed texture testing devices are not utilized because of poor correlations with panel tests of questionable value, or because improper methods were used to correlate the two measurements.

The philosophy and pitfalls of correlating sensory with objective texture evaluation have been discussed recently by Kramer (1969) and Szczesniak (1968). Problems of accuracy and precision, questions of sample range and distribution, and testing methodology were considered.

The purpose of this writing is to examine selectively certain theoretical and practical aspects of mathematical correlations and other types of associations between sensory and objective data and to draw attention to some of the inherent problems in arriving at such relationships.

2. General Considerations

The following factors must be considered in relating instrumental to sensory texture measurements:
- properties measured by the instrument, test conditions and data interpretation.
- physiological, psychological and other factors influencing the panel.
- heterogeneity or time-induced changes in the test material.
- correlating methodology.

A. Kramer and A. S. Szczesniak (eds.), Texture Measurements of Foods, 130–160. All Rights Reserved.
Copyright © 1973 by D. Reidel Publishing Company, Dordrecht-Holland.

Proper selection of an instrument and of test conditions cannot be overemphasized. Factors such as temperature and the presence of moisture must be considered when testing materials which are heat and/or moisture sensitive. This is especially important for materials which show drastic textural changes on being contacted with saliva and/or temperature.

Assuming that the instrument does measure parameters relevant to sensory texture evaiuation, it is imperative that the tested samples represent the commercially important range of intensity of the particular characteristics. It is also essential that the samples be evenly distributed throughout the range, that they cover all pertinent conditions and variations, and that an adequate number be tested. Kramer (1969) suggested that a minimum of 50, and preferably more than 100, samples be tested in studies dealing with correlating instrumental with sensory evaluations.

Since both instrumental and sensory measurements are generally destructive in nature, it is not possible to perform the two types of evaluation on exactly the same sample. Unless an adequate number of replicates is tested, this introduces serious errors if the test material shows significant sample-to-sample variations. Also, since usually there is a time lapse between objective and panel tests, any significant time-induced changes in the samples may contribute to poor correlations.

Methods used in determining the degree of correlation will be the main subject of this chapter. It is not, however, our intention to present a treatise on statistical methods for which the reader is referred to specific publications (e.g. Coombs, 1964; Fisher, 1969; Snedecor and Cochran, 1967; Sokal and Rohlf, 1969; Wine, 1964; also ASTM, 1967; Kramer and Twigg, 1970; Tarver and Schenck, 1958).

3. Methods for Expressing Relationships

Associations between subjective and objective texture measurements may be expressed in graphical or mathematical-statistical terms. In a study of a possible relationship between two types of variables, a plot of points on an x-y coordinate system is usually the first step; clues on linearity, necessity of fitting a curve of higher order, patternless scatter, or even some fault in the collection of the experimental data may be gained by an inspection of this plot.

The presence of linearity greatly simplifies the study of a relationship and the making of possible predictions. Because of this advantage, even if linearity is not at first evident, transformations of data are often used to achieve a linear relationship. Almost any curve can approximate a straight line when sufficiently short intervals are considered. Thus, when the range of texture values of commercial interest is narrow, this treatment may be useful. However, such an approximation may obliterate important differences which need to be investigated.

If a linear relationship is not at first evident, conversion of one or both variables into a different form may result in a straight line. Such transformations may involve (a) expression of the instrumental data in a different physical or (b) mathematical form. Examples of the first category are use of work instead of compressive force, degrees of an angle instead of percent. In studies of the degree of coolness in the mouth of whipped cream-like products, Szczesniak (1968) discussed a case where a calorimeter was used to measure objectively the amount of heat absorbed within the approximate temperature range to which the sample was subjected in the mouth. When calorimeter results were expressed per gram of material, the coefficient of linear correlation was 0.62, which was significant at the 5% level; when the same data were expressed per 5 ml (volume of material used for sensory evaluation), a coefficient of 0.94 significant at the 0.1% level was obtained. As evidenced by this example, physical transformation may lead also to an increase in the correlation coefficient beyond that obtained by improving the linearity of the relationship.

Table I shows a number of different types of mathematical linearizing transformations of which logarithmic conversions are the most common. Many nonlinear

TABLE I

Some linearizing transformations (Natrella, 1963)

No.	Original relation	Transformed variables $Y_T =$	$X_T =$	Straight line $Y_T = b_0 + b_1 X_T$	Conversion of straight line constants (b_0 and b_1) to original constants	
					$b_0 =$	$b_1 =$
1	$Y = a + b/X$	Y	$1/X$		a	b
2	$Y = 1/(a + bX)$ or $1/Y = a + bX$	$1/Y$	X		a	b
3	$Y = X/(a + bX)$	X/Y	X	In all formulas given, substitute values of Y_T for Y and values of X_T for X, as appropriate.	a	b
4	$Y = aX^b$	$\log Y$	$\log X$		$\log a$	b
5	$Y = a + bX^n$, where n is known	Y	X^n		a	b
6	$Y = ab^x$	$\log Y$	X		$\log a$	$\log b$
7	$Y = ae^{bx}$	$\log Y$	X		$\log a$	$b \log e$

subjective-objective relations can be converted into a linear form when one or both variables are expressed in logarithmic terms. This is especially true if, e.g., the change in the property is proportional to its size (exponential relationship). Skewed curves may be converted into normal distribution curves when logarithms instead of the original numbers are used, but the validity of this treatment should always be substantiated by an actual knowledge of the experimental material. One quick method for determining the potential for improving linearity is to compute the correlation coefficient by a parametric (r) and non-parametric (ϱ) procedure. Where ϱ is significantly higher than r, curve fitting is indicated (Kramer, 1969).

Table II presents four types of linear relations when two variables are considered

TABLE II
Linear relationships

	Statistical		Functional	
	S1	S2	F1	F2
Fitting line	$\bar{Y}_x = c_0 + c_1 X$	$\bar{Y}_x = c_0 + c_1 X$	$Y = c_0 + c_1 X$	$X = c_0 + c_1 \mathscr{F}(P)$ $Y = d_0 + d_1 \mathscr{F}(P)$
Restrictions	Both X and Y are measured on samples selected at random	X is fixed by choice Y corresponds to (depends on) the value of X	X accurately known Y normally distributed	$f(P)$ may be linear, quadratic, exponential etc. Response curves of X and Y are linearly related
Errors of measurement	Negligible compared to variation among samples	Same as S1	Measurement error affects Y only	X and Y are both subject to error
Correlation coefficient	Sample estimate is $r = \dfrac{S_{xy}}{\sqrt{S_{xx}} \cdot \sqrt{S_{yy}}}$	Correlation may be present, but r calculated from this type of experiment will give a distorted estimate of the correlation	Not applicable	Not applicable
Example	X = Instron measurement of ultimate strength Y = sensory measurement of 'hardness'; both X and Y are obtained on samples randomly selected. (X is not selected but it 'comes with' the sample unit)	X = instrumental viscosity. Control of sample concentration makes possible preselection of a certain value of X beforehand. Y = sensory 'consistency' corresponding to the preselected value of X	X = accurately known weight suspended from a spring which obeys Hooke's law Y = corresponding elongation of spring	X = reading of weighing balance #1 Y = reading of weighing balance #2 P = weight

(Natrella 1963). For a detailed treatment of underlying assumptions, type of predictions possible, and applicability to different experimental situations, the reader is referred to the work by Acton (1959). Only brief comments on main points, as applied to texture work, will be made here.

In all functional relationships, it is assumed that an exact mathematical formula is present relating Y as a function of X. The only reason that the experimental data may not fit the equation is because of errors of measurement in one or both variables. With the possible exception of some psychophysics laws, most (if not all) relations between subjective and objective texture measurements are statistical rather than exact mathematical functions between X and Y, where Y, the dependent variable, represents the psychological (sensory) value, and X represents the independent, physical value or values from which Y may be predicted. Where relations cannot be made to approximate linearity even through transformations, polynomial techniques are employed for the computation of the least-squares fitted curve.

4. Correlation and Regression

4.1. SCATTER DIAGRAMS

Before a statistical analysis is applied, plotting of the data in the form of scatter diagrams is helpful in revealing important information on the relationship between variables. Examples of scatter diagrams relating sensory consistency of applesauce to six different objective tests and adopted from Kramer and Twigg (1970) are shown in Figure 1. Visual examination readily indicates that of the six tests used, only two (D and F) hold promise of a high correlation with panel scores. Before a correlation analysis is applied, these authors suggest the following useful procedure. Each scatter diagram is divided into four quadrants and the number of points in quadrants I and III (direction of a positive correlation) and in quadrants II and IV (direction of a negative correlation) are added separately. Considering scatter diagram A as an example, quadrants I and III have a total of 6 points and quadrants II and IV a total of 4 points. Statistically, when the total number of paired comparisons (in all quadrants) is 10, a minimum of 9 points is required in order to obtain a significant correlation at the 5% level. (In practice this information can be readily obtained on the basis of a table relating the total number of paired comparisons with the number of correct choices required for the different levels of significance). Since in this example the number of points in quadrants I and III is only 6, one can assume that no statistically significant positive correlation between the panel scores and the objective test used in A was demonstrated. Similarly, since the number of points in quadrants II and IV is 4, no significant negative correlation was shown. In contrast, examination of scatter diagrams D and F shows that quadrants I and III have enough points for a significant positive correlation (10 each).

Subsequent to plotting of the data, statistical analysis can provide probability-based quantitative information on the relationship between objective and subjective measurements.

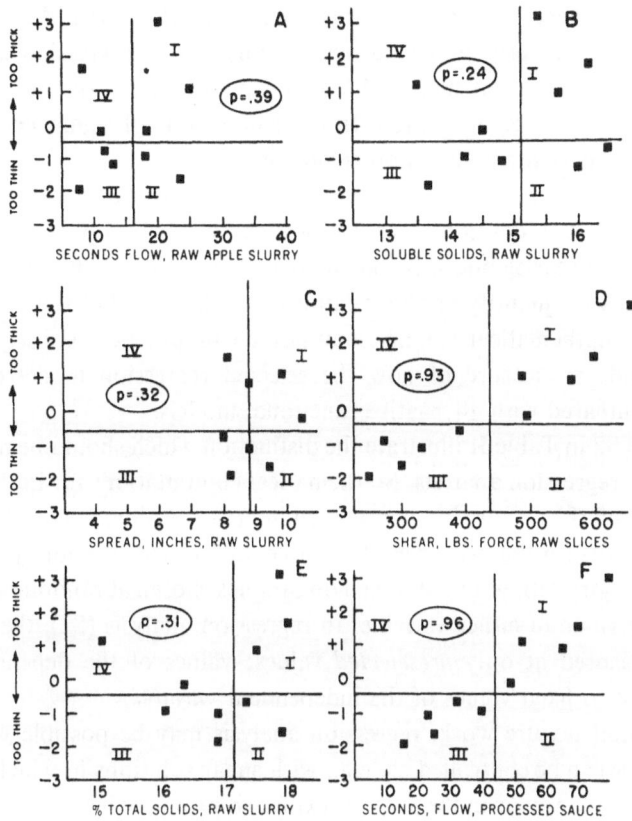

Fig. 1.   Scatter diagrams indicating the degree and nature of correlation and regression of various objective tests to sensory evaluations (Kramer and Twigg, 1970).

4.2. Correlation vs regression analysis

It is important to note the basic distinction between correlation and regression analysis. In correlation analysis, the basic question is whether or not two variables move together. There is no assumption of causality. In fact, it may be that changes in the two variables may be the result of a third variable which may be unspecified. For example, there is a high correlation between the consumption of wine and the salary of teachers in the United States. Instead of concluding that the moral fiber of teachers is deteriorating, the usual explanation is that the rising levels of both are the result of rising per capita income in the United States.

In contrast, regression analysis has implicit in it the assumption of a unilateral causality. That is, changes in X result in changes in Y. However, changes in Y do not result in changes in X. This unilateral causality is basic to the mathematics underlying regression analysis. For example, a regression where the height of the son (Y) is a function of the height of the father (X) is consistent with the basic assumption of a unilateral causality. Texture measurements and the relationship between sensory and objective measurements are considerably more complex than the mathematical assumptions which underlie either correlation analysis or regression analysis. This does not mean that these statistical techniques cannot be used. Instead, it means that they must be used with great caution. In particular, tests of significance will not be valid when the underlying assumption is violated.

Since one of the primary reasons for estimating the relationship between sensory and objective data is to develop objective means of evaluating products in a manner correlatable with organoleptic assessment, regression still can be useful because it provides a means of quantifying this relationship. The fact that tests of significance have lost their mathematical validity may not be of practical consequence. These tests still provide a measure of how the selected regression model describes the relationship compared with alternative functional models.

Cases S1 and S2 in Table II illustrate the distinction which should be made between correlation and regression analysis. Mathematical computations for both are, in most part, common and the two are frequently used interchangeably. In correlation analysis (S1), measurements for the two types of variables are selected randomly: instrumental readings and sensory ratings are obtained on samples chosen at random, i.e. regardless of the absolute value of either variable. In regression analysis (S2), the independent variable is measured at only *preselected* values; values of the dependent variable correspond then to fixed values of the independent variable.

In experimental texture work, regression analysis may be possible when the mechanical variable can be controlled as, e.g., with sugar solutions in which the concentration governs viscosity or in the case of texturized protein where certain mechanical characteristics can be controlled by processing and other means. However, in the majority of cases, selection or fixing of one variable is not possible and, therefore, use of randomly obtained data leads to correlation analysis.

The ideal case of a correlation analysis would be the statistical association between two nondestructive measurements; for example, a random sample is drawn from a population and the weights and thicknesses are measured. The error involved in measuring these two variables is small compared to the variability among samples. In contrast to functional relationships where the presence of an exact mathematical function is assumed, in this example it will not be possible to predict the exact thickness of a sample from its weight, but it may be possible to estimate the average thickness of all samples of a given weight.

4.3. IMPROVING THE CORRELATION

Any calculated correlation coefficient (r) must have a value between 1.0 (or -1.0) and 0.0, with 1.0 indicating a perfect relationship between the two variables and 0.0 indicating none.

The magnitude will depend on several factors, including the type of tests, conditions under which they were applied, kind of food, processing method, and preparation (temperature of serving, etc.). These conditions should be stated explicitly in the publication of results, if reports from different laboratories are to be compared. The lack of attention in this area may, at least partly, explain the wide variation in reported values for the same test on the same type of food.

Correlations reported between Warner Bratzler shear and sensory tenderness of beef vary between -0.16 and -0.94, depending on the type of cut and several other factors (Szczesniak, 1968).

Henry et al. (1971), in a study of several sensory attributes of texture and mechanical tests, using the Instron on the basis of the General Foods texture profile analyses, reported correlation coefficients varying from -0.48 between sensory smoothness and mechanical firmness, to 0.90 between sensory stringiness and the mechanical value of the 'extension of the material at cleavage.'

A recent example of the application of correlation coefficient and analysis of variance in texture measurements is the work of Toda et al. (1971) which refers to a 'principal component analysis' of instrumental and sensory texture descriptions of several food gels and pastes. Over 80% of the total variance could be explained by three components. The first component corresponded to parameters measuring breaking stress or 'hardness', the second to parameters measuring 'springiness', and the third to parameters measuring 'adhesion'.

It is not only valid but desirable to approach the 'real' correlation by reducing errors caused by inadequate sampling, or lack of precision in the performance of the panel or instrumental measurement. Since both subjective and objective texture measurements are generally destructive, separate samples must be used for the two methods. Due to the large variability within the experimental material, the sample tested by an instrument frequently is not a 'true replicate' of the sample evaluated by sensory means. For this reason, considerations such as the number of samples to be tested, the method of preparation, and the techniques of sampling and randomization are of critical importance.

Generally, lack of precision is ascribed to the subjective rather than the objective textural measurements. Furthermore, since a correlation coefficient (r) between a set of panel scores and objective measurements cannot possibly be significantly higher than a coefficient (r_1) between two panels for the same set of samples, it has been suggested (Guilford, 1965) that error assignable to panel variance be corrected by dividing r by $\sqrt{r_1}$. For example, if the r value between panel scores and objective measurements is 0.90, and r_1, between the two panels is only 0.81, then $0.90/\sqrt{0.81} = 1.0$, i.e. the 'true' correlation between sensory and objective evaluation.

Figure 2 is an example of an incorrect 'improvement' of r, illustrating the influence of the range of measurements on the magnitude of the correlation; the risk is evident of drawing conclusions on the basis of a correlation coefficient which has been artificially boosted by the inclusion of two extreme points. Similarly, where sample selection is limited to textural quality levels clustering just above and below one minimum grade point, the resultant correlation is misleadingly low.

The inclusion of extreme values can result also in deviation from linearity due to the psychological tendency of panelists to avoid extremes, i.e. to downgrade the

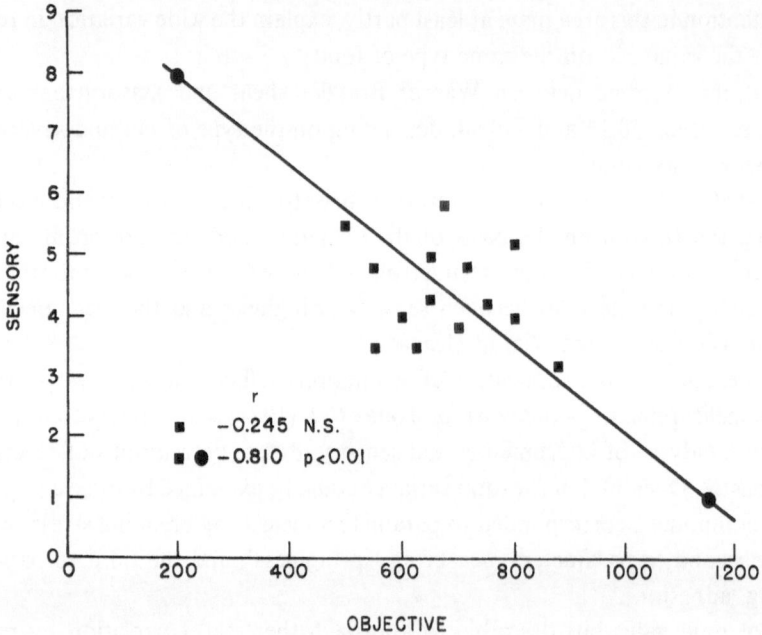

Fig. 2.   Effect of texture range on significance of correlations (Szczesniak, 1968).

'perfect' sample and upgrade the poorest product. In such cases, the original curve may be distorted into a sigmoid (cubic) shape. The value of r^2 rather than r should be used for predicting (or explaining) Y (or X) in terms of X (or Y). Thus if $r_{y \cdot x} = 0.90$, it may be inferred that variations in Y may be explained by up to 0.81, or 81% in terms of X, since $0.90^2 = 0.81$.

4.4. RANK CORRELATION

A simplified quick method of correlation analysis, which has the added advantage of being nonparametric (i.e. normality of distribution need not be assumed), is the rank correlation. In this method, the subjective scores and the objective measurements are converted into rank numbers. The rank numbers of one variable (e.g. subjective scores) are arranged in ascending or descending order against the corresponding rank numbers of the other variable (ex. objective measurements). The differences between these paired ranks are then used to calculate the rank correlation coefficient as follows:

$$\varrho = 1 - \left[\frac{6 \sum (\text{R.D.})^2}{n \cdot (n^2 - \perp)}\right], \tag{1}$$

where R.D. = difference between paired ranks, n = number of pairs, and $\sum$ = summation over n values of $(\text{R.D.})^2$. Litchfield and Wilcoxon (1955) developed a nomogram of n^2, ϱ and $\sum (\text{R.D.})^2$ scales for the quick solution of this equation.

5. Multiple Regression

Multiple regression is a statistical case of multivariate association where a dependent variable is influenced by two or more variables as in the equation:

$$Y = a + b_1 X_1 + b_2 X_2 + \cdots b_n X_n, \tag{2}$$

where Y represents the dependent variable and $X_1, X_2 \ldots X_n$ represent several independent variables. Coefficients b_1 and b_2 are termed *net* regression coefficients to indicate that they show the relation of Y to X_1 and X_2, respectively, excluding or *net of* the associated effects of the other independent variable(s). Thus, multiple regression is concerned with the relation of the dependent variable to each independent variable, as well as with the composite relation of the dependent variable to all the independent variables together.

In texture work, considering a case of two independent variables, Y in Equation (2) may represent the sensory tenderness of meat, X_1 may be the modulus of elasticity and X_2 the ultimate strength. The constant b_1 is termed 'the net regression of Y on X_1, holding X_2 constant'. Stated otherwise, b_1 refers to the average change observed in tenderness Y with unit changes in the modulus of elasticity X_1, determined while simultaneously eliminating from tenderness Y any variation accompanying changes in the ultimate strength X_2. Application of multiple regression analysis to these data will provide information on (a) how closely values of sensory tenderness, estimated on the basis of the regression equation, could be expected to agree with the actual experimental values; and (b) what proportion of the total observed variation in tenderness could be explained or accounted for by its relation to the modulus of elasticity and the ultimate strength. Question (a) will be answered by the standard error of estimate, while question (b) will be dealt with by the coefficient of determination (R^2). Thus, multiple regression analysis presents in a condensed form a large amount of information superior to the estimates obtained by linear equations of any other form.

An example will serve to illustrate the practical application of this type of mathematical computation. Rasekh *et al.* (1970) used single and multiple regression analysis in the study of canned tuna to arrive at a relationship between consumer panel acceptability as the dependent variable and different objective tests as independent variables. On the basis of non-parametric rank procedures, they obtained the highest single correlation coefficient with consumer data using the Hunter b value which is a measure of yellowness ($r = 0.559$); this was followed by the maximum shear force value ($r = 0.558$), the Hunter L value ($r = 0.532$), the pH ($r = 0.398$) and the percent fiber ($r = 0.289$). The *multiple* correlation coefficient (R) between consumer panel and these 5 objective tests was 0.791. On the basis of the *partial* regression values obtained for each of the independent variables, the contribution to the consumer score (C) of the Hunter L test was approximately 10%; the Hunter b test, 25%; the shear test (S), 30%; the fiber test (F), 20%; and the pH test (P), 15%. These results were incorporated into the following equation:

$$C = 4.918 - 0.0243L + 0.224b + 0.00346S - 0.0706F + 0.804P.$$

This allows to predict consumer acceptance of another sample of canned tuna following its characterization by objective tests. If this sample has a Hunter L value of 50, a Hunter b value of 10, a shear value of 700, percent fiber of 5, and pH of 6.4, the corresponding consumer panel preference score, calculated on the basis of the multiple regression equation, would be 3.32, which is classified as a 'medium' grade.

A typical example of the application of multiple regression analysis to texture data obtained by physical and sensory testing is the work of Henry *et al.* (1971) on a series of semi-solid foods. Four factors identified for each set of data accounted for more than 90% of the variations in each case. Multiple regression equations were established for predicting the four factors from the sensory tests by means of the physical test values.

For further applications of multiple regression analysis to foods, see Harper and Baron (1951); Fox and Kramer (1966); and Yoshida (1968). For a lucid exposition of the principle of linear and non-linear regression analysis and for methods of fitting the different curves see Steel and Torrie (1960); Ezekiel (1959) and Kramer and Twigg (1970).

6. Non-Linear Relationships

Since many relationships between objective and subjective texture measurements are non-linear, non-parametric ranking procedures may be used as a first approach to the treatment of the data. The next step may then be to determine the exact curvilinear relation which describes the extent to which changes in the dependent variable are associated with changes in each particular independent variable, while simultaneously removing that part of the variation (in the dependent variable) which is associated with the other independent variables. Estimating such relationships requires no specialized procedure since any curvilinear function can be visualized as a multiple regression problem and estimated as such. Thus, for example, the curvilinear function:

$$Y = a + b_1X_1 + b_2X^2 + \cdots b_nX^n \tag{3}$$

could be treated as a multiple regression problem where $X_1 = X$, $X_2 = X^2$, and $X_n = X^n$. The new equation is the same as the multiple regression Equation (2). Thus, a relationship visualized as curvilinear in a two-dimensional world (X, Y) becomes linear in a three-dimensional world (X, X^2, X^n). Adding higher powers to a polynomial simply results in adding more dimensions to the multiple regression model. These 'functional' polynomials ($F1$ and $F2$) are shown in Table II as the linear analogs of $S1$ and $S2$.

The fitting of polynomials is the commonest technique for dealing with non-linear relations. Within this technique, the use of *orthogonal polynomials* allows the computation of the least-squares fitted curve of any degree. For the explanation and use of this procedure, the reader is referred to the work by Acton (1959), and Kramer and Twigg (1970).

In texture work, the applicability of functional relations is limited mainly to mechanical measurements where exact mathematical relationships can be shown to exist; the development of mechanical models is an example. In contrast, most work on relationships between subjective and objective texture measurements belongs to the statistical type of association. Within this class, analogs of the $S1$ and $S2$ cases of Table II are of the form:

$$\bar{Y}_{(x)} = f_0 + f X_1 + f_a X_2 + \cdots + f_{k-1} X_{k-1}, \tag{4}$$

where $\bar{Y}_{(x)}$ is the mean value of all the Y values which are derived from the combination of $X_1, X_2 \ldots X_{k-1}$ values, and where

$$f_0 = M_y - f_1 M_{x1} - f_a M_{x2} - \cdots - f_{k-1} M_{Xk-1}. \tag{5}$$

The letters $M_y, M_{x1} \ldots, M_{xk-1}$ indicate the population means of $Y_1, X_1 \ldots X_{k-1}$, respectively.

In all the above cases, the relationships between Y's and X's is linear with regard to the coefficients, and the task is to determine the values of these coefficients on the basis of the experimental data. The general method for achieving this goal and for answering questions about the relationship as a whole (equation for prediction, estimates of confidence intervals, etc.) is the *method of Least Squares*. For the theory and the particular procedure followed, especially for the arrangements and steps leading to a form suitable for automatic computation by computers, the reader is referred to the available literature (e.g. Davies, 1954; Dwyer, 1951). A simple, ruler-and-pencil graphical method applying to data which are equally spaced in the X dimension has been described by Askovitz (1957).

7. Sources of Variation

In considering a correlation analysis of instrumental and sensory texture data, three sources of variation are important. The first is the *primary source* of variation which the experimental work strives to measure and correlate. Instrumentally, this may be shear force values obtained on different meat samples; sensorily, this may be panel ratings for tenderness of replicate samples. If a correlation exists between shear force values and tenderness ratings, the two types of data will be associated; they will tend to co-vary due to common cause(s) which operates on both the mechanical and the sensory measurements. The second source of variation is the *random errors* of instrumental and sensory measurements. These are assumed to be non-existent, or normally and independently distributed. The third source is the *extraneous sources* of variation. In an instrumental method, they may include inaccurate calibration, differences due to instrument and operator performance at different times, and effects due to the heterogenity of food samples. In a sensory method, they may include differences in motivation, sensitivity, and repeatability of panel members. The random errors and extraneous sources of variation can be considered 'alien' to the primary source of interest. Extraneous effects tend to decrease the magnitude of or obscure an existing correlation. These effects are often inseparable among themselves, as well as inseparable from the primary sources under study.

Approaches which have been used to remove variations due to extraneous effects (aside from improvement of sampling procedures and better control of experimental conditions) include: (a) the method of fitting constants, (b) transformation of the experimental material, and (c) methods of data analysis.

7.1. METHODS OF FITTING CONSTANTS

This involves the use of additive or multiplicative corrections on the observations in cases where the specific extraneous effects can be separated. If for example, temperature and humidity affects the strain gauge response of a particular instrument in a predetermined manner, corrections can be applied to the individual measurements. This is another way of stating that test procedures should be properly standardized and calibrated.

7.2. MODIFICATION OF THE EXPERIMENTAL MATERIAL

Suppose the task is to study the correlation between mechanical texture measurements and sensory ratings when the two types of measurement apply to a highly variable experimental material, e.g., meat. Due to the heterogeneity which is encountered, even between adjacent locations within a single muscle, the characteristics of samples subjected to mechanical and sensory testing may vary widely. One approach is the use of an adequate number of samples, randomization, systematic distribution, and other techniques necessitating a large investment in time and effort, in order to ascertain the possible covariation of the two types of data discussed above. A different technique is to deliberately achieve uniformity by transformation of the experimental material, e.g., by grinding and mixing of the meat throughout the muscle; this will assure true replicates between mechanical and sensory measurements. Assuming such treatments affect all lots similarly, ground lots of meat from muscles of different textural quality may be tested, and the correlation between mechanical and sensory measurements may be studied over a range of values.

Anderson and co-workers (1972a, b) used such an approach in an effort to (a) study the correlation between mechanical measurements and sensory ratings of tenderness on cooked beef, and (b) express the mechanical-textural qualities of a whole carcass by a single number. The probability that a certain mechanical or sensory measurement on modified or 'processed' muscle represents a physical average of the tenderness values of the individual parts will depend on the technique (grinding, mixing, etc.) used to modify the muscle. The technique also assumed that the structure of the test material does not affect its measured characteristics, or that the effect may be disregarded in view of the pre-set objectives of the work. To achieve the goal of expressing the textural qualities of a whole carcass by a single number, these researchers suggested that the mechanical tenderness readings of individual muscles be 'weighted' according to their relative contribution either to the total weight or to the total value. It should be noted that such 'weighting' is equivalent to the partial regression coefficients (b) in Equation (2).

7.3. DATA ANALYSIS

A method of data analysis for removing extraneous sources of variation has been applied recently by Gacula *et al.* (1971) in correlating Warner-Bratzler shear measurements and sensory ratings obtained by a trained panel on cooked beef muscles.

Individual observations were expressed *as a function of the contemporary* mean, defined as a mean computed from observations correlated in the same substratum which was assumed to be homogeneous by proper blocking, grouping, and balancing. Four substrata (subpopulations) were established representing four states of experimental runs taken at monthly intervals. This subgrouping was done on the basis of personal assessment of the background conditions during the collection of data. Finally, in place of the actual data, deviations from the mean of the substrate were considered. When the observations were expressed in this way, the data were free of the extraneous sources of variation. The results of this analysis can be seen by comparing the middle graph of Figure 3 which contains the actual data with the bottom graph which is based on deviation data from the contemporary mean. The low correlation of -0.36 in the middle plot was traced to external influences affecting the subpopulation in a different manner, and to the instability of the population means upon time, especially in the taste panel data. All these effects tend to mask the real correlation. The deviation plot (bottom part of Figure 3) resulted in a correlation coefficient of -0.86 which is a substantial improvement over that obtained by using the original data.

A more elegant method for achieving similar results is the Analysis of Variance (Fisher, 1969) used extensively in agricultural and industrial research for half a century. In sensory work, it is the 'validation' step in the procedure for developing an objective method proposed by Kramer (1973). Specifically, the deviation technique has been used by Smith and Kramer (1972) to improve the relations between nutritive value and consumer acceptance of prepared collard greens processed by various methods.

7.4. CORRECTING FOR PANEL VARIABILITY

As stated previously the lack of precision in subjective evaluation is an important primary source of variation. It may be corrected by dividing r by $\sqrt{r_1}$ where $r =$ correlation between panel and objective data, and $r_1 =$ correlation between 2 panels.

8. Prediction

As stated earlier, except for functional relationships where the different variables are related by an exact mathematical formula, in texture work prediction of one variable on the basis of values for the other variable(s) depends on statistical methods.

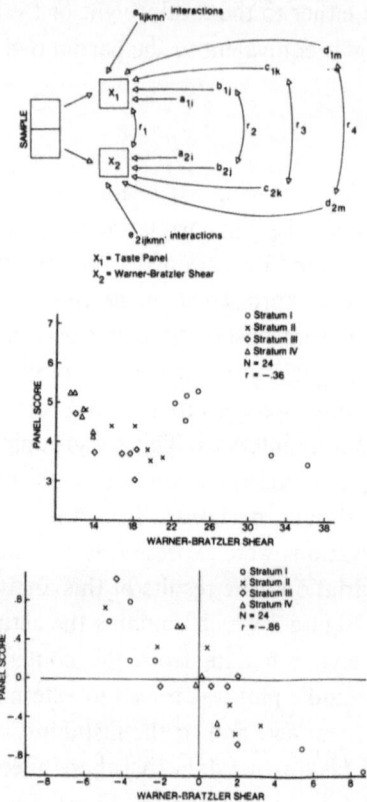

Fig. 3.   Correlations between objective and subjective texture measurements (Gacula *et al.*, 1971). – *Top:* Schematic representation of correlated causes between sensory panel and Warner-Bratzler shear. – *Middle:* Scatter diagram of raw data. – *Bottom:* Scatter diagram of the same data when expressed as deviations from their contemporary mean.

8.1. CORRELATION AND REGRESSION ANALYSIS

Correlation analysis ($S1$ in Table II) presents a unique case where two types of prediction may be made: if $0 < r < 1$, X may be predicted from Y on the basis of the regression line $\bar{X}_y = C_0' + C_1' \, Y$ and Y may be predicted from X on the basis of the regression line $\bar{Y}_x = C_0 + C_1 \, X$ (Figure 4). In regression analysis ($S2$ in Table II), there is only one type of prediction: if mechanical measurements were made at preselected values, the only prediction possible is that of sensory ratings; if sensory ratings could be obtained at preselected values, the only prediction possible would be that of mechanical measurements. Once an experiment has been designed on the basis of the $S2$ case, and a regression line was fitted to the equation $Y = C_0 + C_1 \, X$, where X is the variable measured at preselected intervals, the use of the same data to fit equation $X = C_0' + C_1' \, Y$, will not provide an unbiased estimate of the true regression line in the unrestricted population. If a 'prediction' were attempted under these conditions, gross distortion could result.

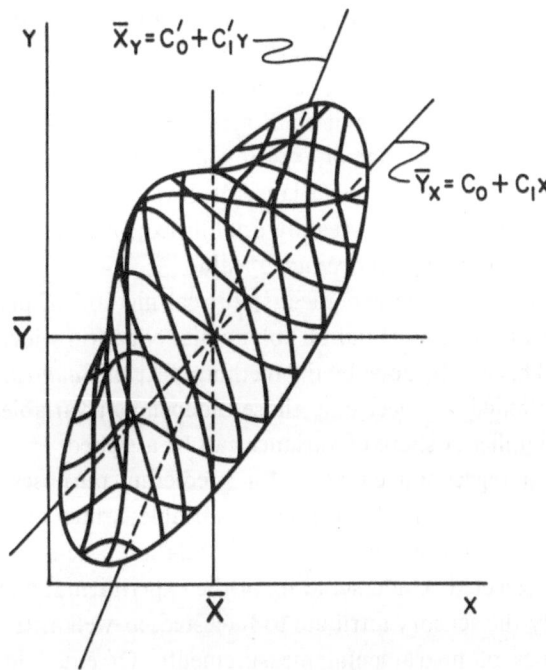

Fig. 4.   Normal bivariate frequency surface.

One of the practical advantages of a regression line is the use of confidence limits. In proceeding with this method, experimental values for the independent (fixed, selectable) variable X are plotted on one ordinate, and the corresponding experimental values for the dependent variable Y are plotted on the other ordinate. At each level of X, the variance of Y values is determined. If all such variances are the same (or relatively constant), confidence levels at specified levels of significance can be placed on both sides of the regression line.

In order to establish confidence limits, it is necessary to compute initially the Standard Error of Estimate $(Sy \cdot x)$ as follows:

$$Sy \cdot x = \sqrt{\frac{\sum (Y - \bar{Y})^2}{n - 2}} \tag{6}$$

where $(Y - \bar{Y})$ measures for each plotted data point the difference between the actual Y and its respective regression line value $\bar{Y}$, and n is the total number of data points (sample size). This statistic is a measure of the dispersion of Y values and when squared, it is an unbiased estimate of the assumed constant variance for the normal distribution of Y about the regression line. The $Sy \cdot x$ quantity is multiplied by an appropriate 't' value. The latter statistic is obtained from 't' tables and its exact value will depend on the confidence interval (90, 95 etc. %) desired, and on the degrees of freedom $(n-2)$. The result of the multiplication is then added and subtracted from the constant C_0 of the regression line $(Y = C_0 + C_1 X)$ thus obtaining the upper and lower confidence limit lines.

This will indicate that the value of Y, i.e. sensory rating, which was calculated from the regression equation on the basis of an X measurement, i.e. mechanical measurement, will fall within a certain range of experimentally determined sensory ratings at a specified level of probability. If the variance of Y values is not constant along the X axis, confidence limits cannot be readily established, but the regression equation may still be useful in predicting an *average* Y value.

In multivariate analysis, a mathematical-statistical method of prediction depends on the development of an exponential or polynomial equation which will best fit the experimental data. This can be done by the method of *Least Squares*, where computer techniques can be applied to select only those independent variables (and/or power terms) to which a significant share of variance can be assigned.

The accuracy of a regression equation for predicting purposes depends on the proper selection of the mechanical and sensory variables which are to be correlated. Computer techniques cannot produce the high correlation necessary for prediction, if wrong variables are selected. Understanding of the experimental material is essential for defining precisely the sensory attribute to be tested, as well as the selection of the most promising types of instrumental measurements. Of equal importance is the mobilization of an adequate number of samples representing the entire range of quality of the sensory parameter to be predicted. If these conditions are not met, the treatment of the data becomes a purely mathematical exercise, and the analysis may well include: (a) development of high order, cumbersome equations, that may be difficult to solve and use, (b) lack of physical significance in the various terms of the equation, and (c) absence of sufficiently narrow confidence limits for probability prediction. An equation could be developed to fit almost any sets of values, but the reliability of prediction when samples differing from the sampled population are considered, may be quite low.

8.2. RESPONSE SURFACES METHOD

In order to select the most promising or 'best' variables, the applicability of the method of *response surfaces* may be worth investigating. In this method, fixed levels of one variable are tested against varying levels of the other(s), the response surface assuming the form of a topographic map. In a three-dimensional pattern, optimum conditions are shown on points of a surface of a solid. Although additional dimensions may be calculated, graphical portrayal is difficult. In recent years, great impetus to the application of this approach in process engineering and other disciplines has been given by the use of computers. Although, to our knowledge, no use of this method has yet been reported in problems of objective-subjective texture correlations, this and related approaches might make it possible to construct mathematical prototype models *in advance* of development work. For example, considering a theoretically ideal meat product on the basis of previous experience with consumer preferences and mechanical measurements, the question to answer may be: What combinations of the modulus of elasticity, ultimate strength, work to rupture, and other physical properties would give a maximum sensory tenderness score? For general information on the method of response surfaces, see Box (1954); for specific applications to problems in the food industry see Kissel (1967), Kramer (1965a), MacDonald and Bly (1966) and Smith and Rose (1963); for general information on the use of factor analysis see Cochran and Cox (1957); for applications in foods see Harper (1956), Harper and Baron (1948, 1951), and Henry *et al.* (1971).

9. Analytical, Integrative and Sensory Texture 'Interfaces'

The problem of which mechanical and sensory texture variables should be considered in the planning of an experiment is outside the scope of this discussion. The reader is referred to the books by Mohsenin (1970a) and Sherman (1970) with regard to mechanical properties and their measurement, and to publications by Amerine *et al.* (1965), Kramer (1964), Kramer and Twigg (1970), LeMagnen (1962), Sherman (1969), Szczesniak (1963b), Szczesniak and Kleyn (1963), Yoshida (1968), and Yoshikawa *et al.* (1970) with regard to subjective texture profiles, glossaries, and variables.

Table III lists a number of mechanical measurements under analytical (engineering, rheological) and 'integrative' (consumer or product oriented, empirical) categories. From the physical or mechanical points of view, analytical terms are well-defined and more or less universally accepted. In the food field, there is relatively little published work on correlations between rheological measurements and sensory texture properties. This emerging area of research holds great promise to the food industry, and one can expect an increasing number of publications in 'food psychorheology'. However, in spite of the fundamental significance and the challenge of well-defined analytical measurements, integrative approaches to the objective measurement of food texture will continue to have an appeal in correlation work with sensory data. The reasons for this include: (a) simplicity in obtaining a measurement, and (b) relative success in achieving some good correlations with sensory panel data.

From the rheological point of view, an integrative texture property may be extremely complicated. For example, 'chewiness' may include contributing components of rigidity, ultimate strength, and dissipated or stored potential energy, as measured by successive or overlapping modes of multiaxial compression, tension, and shear. In spite of this theoretical complexity, integrative tests have been proven valuable in product development and quality control. Therefore, once an integrative method of high correlatability with sensory ratings has been developed, it becomes the challenge of the rheologist through the analytical and reductive approach to interpret 'how and why the method works'. This challenge applies to both the rheological analysis of the empirical device and to the mechanical and psychophysical analysis of chewing. Even with the aid of computers, this appears to be a monumental task.

The arguments of different schools of thought in connection with the analytical-integrative-sensory texture interfaces originate from the fundamental difference between what an instrument actually measures and what a human subject perceives

TABLE III

List of objective texture measurements for correlations with sensory data

---

Analytical Properties (Mohsenin, 1970; Sherman, 1970)

From force-deformation curves at constant rate of strain

Modulus of elasticity
  apparent Young
  initial tangent
  tangent
  secant

Stress
  at given strain
  to yield
  to rupture (ultimate strength)

Work
  at given strain
  to yield
  to rupture ('toughness'; Finney, 1969)

Degree of recovery

Mechanical hysteresis

Frictional energy loss ('crushability index'; Drake, 1966)

Decay (relaxation) of:
  elastic modulus
  stress   *value at ith loading cycle*
  work   value at 1st loading cycle
  hysteresis   (Segars and Kapsalis 1971)

From dynamic testing (Drake, 1962; Finney, 1971)
  complex modulus
  apparent modulus
  loss (storage) modulus
  mechanical damping
  viscosity
  resonance frequency

From stress relaxation and creep curves (mechanical models; Mohsenin, 1970; Sherman, 1970)
  elastic modulus $E_1, E_2 \ldots E_i$ (spring constants)
    compliances $D_i = 1/E_i$
  viscosities $n_1, n_2 \ldots n_i$ (viscous constants)
  relaxation (retardation) times N_i/E_i
  modulus function $E(t)$
  compliance function $D(t) = 1/E(t)$

From bending measurements (Kapsalis *et al.*, 1972)
  modulus of elasticity, E
  bending rigidity, EI, ($I =$ moment of inertia)
  bending moment loss, $2M_f$
  $EI/2M_f$
  curvature set
  critical curvature

Integrative properties[a]

hardness	Kramer (1961), Szczesniak *et al.* (1963b)
cohesiveness	Szczesniak *et al.* (1963b)
adhesiveness	Szczesniak *et al.* (1963b)
springiness	Szczesniak *et al.* (1963b)
brittleness	Szczesniak *et al.* (1963b)
chewiness	Szczesniak *et al.* (1963b)
guminess	Szczesniak *et al.* (1963b)
spreadability	Kapsalis *et al.* (1963)
work	Kramer (1964)

---

[a] Structural classification of these terms within the system of proponent authors is discussed by Abbot in Chapter III, p. 17.

(Harper, 1968). This difference was discussed recently by Corey (1970) who suggested that "a foodstuff cannot have texture, only particular mechanical (and other) properties which are involved in producing sensory feelings or texture notes *for the human being during the act of chewing the foodstuff*". For these texture-inducing properties, he suggests the name 'texturogenes'. If the term texture is reserved only for notes perceived during sensory chewing, the dependent and the independent parts of a function correlating objective and subjective measurements can be clarified as follows: dependent variables = sensory texture notes; independent variables = mechanical measurements of texturogenes. As an example, ratings of tenderness and crispness will apply to sensory dependent variables, whereas measurements such as modulus of elasticity obtained by the Instron, cohesiveness obtained by the General Foods Texturometer, and hardness (maximum force) obtained by the Kramer Shear Press will refer to mechanical independent variables.

It should be emphasized that this order of dependent *versus* independent types of variables is only a matter of convenience; in practice it is often desirable to use instrumental measurements in order to estimate or predict consumer ratings. Also, since a subjective response is immensely more complicated than a mechanical property, the sensory rating is considered the quantity to be predicted and accounted for by the multivariate combination of analytical (or integrative) measurements. One may find, for example, that sensory hardness could be statistically predicted by the use of a function which utilizes not only the maximum deforming force but possibly also the slope of the force-deformation curve, the ultimate strength, and other mechanical properties. This is an idea and a possibility worth investigating.

With regard to sensory evaluation, words which apply to textural properties reflect the versatility and nuances of human experience; they are complex, overlapping, numerous, and often difficult to define (for example, mushiness, tenderness, or crunchiness). The ambiguity, however, may not invariably or primarily be with the perceived sensation. Human judges may 'know' what they mean. Depending on linguistic acuity and imagination, they can use a number of synonyms, which, in combination with an operational description of events taking place during chewing, may 'explain' the perception of a certain stimulus. The frustration begins when one attempts to reduce or analyze a descriptive sensory term into a number of physical components which can be measured objectively. The inadequacy here is not in the human subject but in the physical measurement and in the machine.

An additional complication arises from differences in conditions between the chewing of a food and the mechanical testing (Drake, 1968). In the mouth, the variables of heat, saliva, and enzymes subject the food to continuous change. The latter may be related to hydration, displacement of air pockets by liquid, changes in the degree of dispersion and flocculation, changes in pH, chemical degradation, etc. Thus, the mouth operates not only as a 'testing laboratory' but also as a 'processing factory'. It is known from rheological measurements that the above changes may significantly affect mechanical properties. For example, in solutions of certain natural hydrophilic gums, an increase of temperature alone may produce a drastic decrease in viscosity, the interaction approaching a first order rate relation. The human subject measures and integrates sensory chewing perceptions on a material which undergoes continuous transformation. It is as if testing is done on a long series of different samples which are produced not only by the mechanical destruction of the original structure, but also by biochemical conditions in the mouth. From the beginning of chewing to the time of swallowing, a multitude of tests must have been performed, recorded, and evaluated. In contrast, mechanical testing applies usually to the biochemically un-altered state, and if a single loading-unloading cycle is involved, to only one test. Suppose, then, that at different times during the chewing of a food sample aliquot portions were withdrawn and subjected to mechanical testing by an integrative or analytical instrument. What would be the characteristics of the plot of maximum force, modulus of elasticity, ultimate strength, etc. *vs.* time or number of chewing cycle? How will this compare with a plot of the same properties when derived from the mechanical testing of an original ('unchewed') sample through successive loading-unloading cycles?

The above example illustrates the fact that correlations between subjective and objective texture measurements are of an associative, indirect (at times even coinci-dental) nature, reflecting underlying effects which may be operating in the same or different directions on the two sides of the correlation. For this and other (statistical) reasons, correlation and prediction do not imply equality or even equivalency – more important, they do not imply understanding of the mechanisms behind the association.

10. Interaction Between Texture and
Other Quality Attributes

An important consideration which, due to inadequate information is lacking in much of the work on correlation between subjective and objective texture measurements is the influence of other quality attributes on the sensory response. To what extent can appearance, size, and flavor be expected to influence the judgment of textural quality? Also, to what extent can one textural parameter be expected to influence the perception of another textural parameter?

Logically, it may be expected that a consumer panel should be more subject to such extraneous influences than a trained laboratory panel which has been taught to perform in an analytical manner and detach itself from biases. However, even in the latter case extraneous influences cannot be totally eliminated. Physiological effects cannot be erased and psychological effects may still play a role.

Twigg *et al.* (1956) found it necessary to include a measure of kernel size in a 'trimetric' test and in nomograms for predicting the maturity factor in canned and frozen sweet corn. Von Elbe and Johnson (1971) stated in a recent communication that consumer preference scores for texture of peas changed when color differences were present. Flavor preference scores were affected in a similar manner.

Pangborn *et al.* (1973) discussed the potential interactions between taste and con-sistency of liquid media and stated that since the taste and olfactory receptors are located in such close proximity to the cutaneous, kinesthetic, and thermal receptors in the mouth, it may be expected that alteration of the physical state of an oral stimulus could have a differential influence on gustation... conversely, various taste compounds could alter the perceived oral viscosity by changing the rheological behavior of the viscous material or by inducing salivary secretion to dilute the medium.

Observations were made by researchers working with meat texture that juiciness may affect the trained panel's scores for tenderness, especially when expressed as the 'chew count'. This method involves counting the number of chews required to ready the sample for swallowing. Since the latter involves not only physical disintegration but also proper lubrication, a juicy piece of meat may appear to be more tender than a dry piece.

Much needs to be done on elucidating the interplay between textural and non-textural attributes. To a certain extent, progress will depend on gaining a better knowledge of the relationship between the sensory response (textural or non-textural)

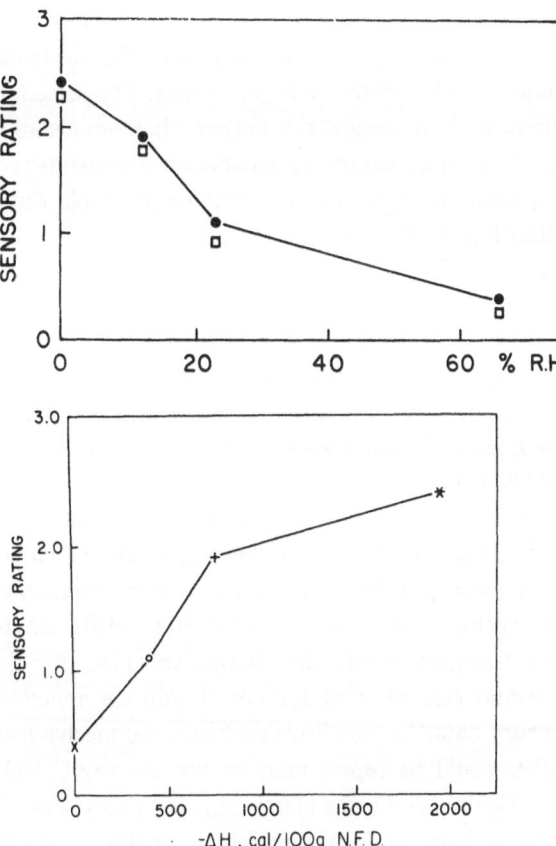

Fig. 5. Relations between sensory heat, perceived by mouth during the chewing of dry meat, and 'objective' heat at different equilibrium relative humidities. Sensory scale: 0 = none, 3 = very evident (Kapsalis *et al.*, 1970). – *Top:* Effect of residual moisture (corresponding R.H.) on sensory heat. – *Bottom:* Sensory heat vs thermodynamic heat content change; * = 0% R.H., + = 12% R.H., 0 = 23% R.H., *x* = 66% R.H.

and other modifying attributes. Some attempts have been made in this area using a thermodynamic approach as illustrated in Figure 5 (Kapsalis *et al.* 1970). The top graph shows heat perceived in the mouth during chewing of dry meat equilibrated at different relative humidities, while the bottom graph shows the relationship between the sensory rating of heat and the enthalpy change (heat content) determined from moisture sorption isotherms for the same test samples. A close agreement between the heat evolved during chewing (due to the moisture pickup by the sample) – a sensory response – and the integral net heat of sorption – a thermodynamic parameter – is interesting and may open the door to a new line of fruitful investigations in the field of psychophysics.

11. Recent Approaches

The problem of correlation between sensory and instrumental data becomes complicated by the fact that rheological properties are rate (and often time) dependent, i.e. physical values obtained with an instrument often depend on the rate at which force is applied to the test specimen. Although the process of chewing also involves different speeds, the conditions are given (fixed) in contrast to the mechanical testing where the optimum rate must be selected by the experimenter. This selection is made difficult by the fact that the rate of chewing varies between individuals and between specific foods and that little basic information is available on stress/strain conditions existing in the mouth. At present none of the instruments available duplicates the force/distance/time relationships in the mouth.

11.1. BUILDING OF MODELS

Figure 6A shows the effect of the rate of testing (cm/min of the Instron punch movement during compression of the sample) on the modulus of elasticity E of graham cracker cubes for different periods of time. The question is: "What speed(s) should be used to study the E value if a correlation analysis with sensory ratings is planned?" One approach would be to use multiple regression in order to examine the contributions of sequentially added testing rates to the magnitude of the correlation coefficient, on a purely statistical-mathematical basis. Another approach, which is meaningful on a rheological basis, is to incorporate all speeds into a mechanical model consisting of elementary units of springs and dashpots (Mohsenin, 1970). In such a model, the elastic constants for the springs and the viscous constants for the dashpots have *physical significance* and can be used for correlation (or regression) studies with mechanical and sensory data. In the above example, the mechanical behavior of the graham cracker cubes could be represented by two Maxwell elements arranged in parallel, as shown in Figure 6B. Of the elastic (E_1, E_2) and viscous (η_1, η_2) constants of this model, the elastic constant E_1 of the first spring showed a trend toward higher values upon storage, as indicated in Figure 6C. This constant then could be used in correlation with sensory work. The relationship, however, ceased to be linear for storage periods in excess of one year at 100 °F. In regression analysis, if the method of magnitude estimation (ME) for sensory hardness were used, the relationship between

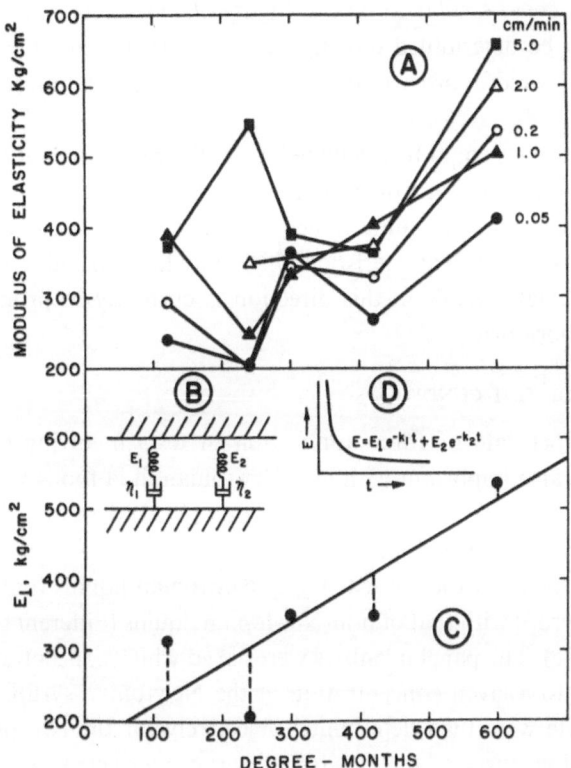

Fig. 6. Application of engineering texture measurements to compressed space cubes in storage (Segars and Kapsalis, 1971). – (A) Effect of the rate of testing on the modulus of elasticity under different temperature-time storage conditions. – (B) Mechanical model calculated from all rates of testing. – (C) Plot of the spring constant E_1 vs temperature-time storage conditions. – (D) Stress relaxation curve and equation corresponding to the mechanical model in B.

sensory and mechanical measurements, according to Stevens (1960), would be:

$$ME = K(P)^M. \tag{7}$$

The exponent M is characteristic of the material; it indicates how rapidly the sensory magnitude grows with increase in the physical property. (See Chapter VIII in this volume, p. 118).

A further step in this type of analysis is the use of the exponential function for the mechanical model:

$$E = E_1 e^{-k_1 t} + E_2 e^{-k_2 t}, \tag{8}$$

where $K_1 = E_1/n_1$ and $K_2 = E_2/n_2$. This equation gives the value of the modulus of elasticity E of the material at any time t in a stress relaxation experiment. In the latter, the material is deformed (compressed) suddenly to a given deformation which is then kept constant while the stress decays. Plotting of the modulus of elasticity E *versus* time t results in a curve similar to that in Figure 6D. (Experimental stress relaxation curves could also be determined directly. In these experiments the use of different testing rates was desired in order to obtain additional information from the recorded curves.)

A number of E values could be calculated from different values of t in Equation (8), especially in the high slope part of the curve in 6D. Substitution of E for P values in Equation (7) will give corresponding values for K and M. These can then be plotted vs t in order to ascertain the characteristics which may have meaning from the psychophysical point of view. Work in this direction is currently in progress at the U.S. Army Natick Laboratories.

11.2. SELECTION OF TEST CONDITIONS

Selecting the proper rate of force application in tests involving correlations with sensory ratings is also important with non-Newtonian fluid foods where the relationship between stress and strain is nonlinear. Wood (1968) determined the appropriate stress rate at which rheological measurements on liquid foods should be made, using subjective comparisons of the consistency of Newtonian liquids (different concentrations of glucose syrup) with that of non-Newtonian liquids (different solid/water ratios of a dry sauce base). The panel members were asked which concentration of the sauce base was nearest to a given concentration of the Newtonian syrup. Curves of shear stress vs shear rate were then determined objectively for the two products, and the point where the flow curve of the Newtonian syrup intersected the curve of the non-Newtonian sauce was taken as the point of equal viscosity. Results for several concentrations of the two test products are summarized in Table IV.

TABLE IV

Comparison of syrup viscosity with a sauce having the same assessed consistency
(Wood, 1968)

Expt No.	Syrup viscosity poise	Shearing stress dyne cm^{-2}	Shear rate s^{-1}	Apparent viscosity, poise n_A
1	10.0	560	56	6.2
2	11.2	506	45	6.4
3	19.4	970	50	8.0
4	19.6	880	45	8.7
5	22.8	1140	50	10.4

It can be seen that (a) the shear rate remained relatively constant while the stress changed with the consistency of the fluid, (b) the apparent viscosity of the sauce did not change linearly with the viscosity of the Newtonian syrup, and (c) in the mouth, within this narrow range of viscosities, the foodstuff is subjected probably to an average shear rate of about 50 s^{-1}, with stress being the perceived stimulus.

In the same work, Wood found a direct relationship between subjective consistency of a soup at different concentrations and shearing stress measured at a shear rate of 50 s^{-1}. The data obeyed the power law $\psi = IS^n$, where ψ is the sensory consistency and S is the shearing stress. The exponent n calculated by the method of least squares was 1.28, with a standard deviation about the regression line of 0.13. In a subsequent experiment, Wood found a linear correlation between the logarithm of the instrumental apparent viscosity of different concentrations of fluids (sauce base and cream of tomato soup) and taste panel scores; instead of magnitude estimation, however, the scale here for the sensory ratings was an equidistant 5-point intensity scale.

Using similar experimental principles, Wood's pionering experiments were extended by Shama and Sherman (1973a) to a variety of products representing a wide range of consistencies, from condensed milk and creamed tomato soup at one end to chocolate spread and glucose syrup at the other. Thirteen different products were evaluated in groups of four by a 26-member panel which was asked to indicate which sample in a pair appeared more viscous in the mouth. Based on matrix analysis of sensory response and on objective flow curves, an estimate was made of the shear stress/shear rate bounds operative in each group of four foods. These are shown as rectangles in Figure 7 superimposed on straight lines which fix the positions of Newtonian viscosities of different values. The double solid curve approximates the region of sensory stimuli operating on the panel. It shows that the stimulus associated with the oral

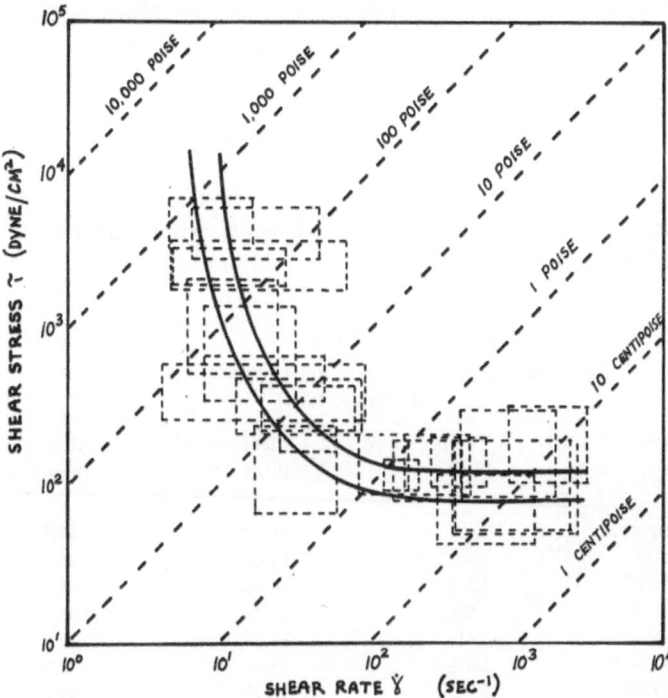

Fig. 7.   Bounds for shear stress and shear rate associated with oral evaluation of viscosity
(Shama and Sherman, 1973a).

evaluation of fluid foods may be the shear rate at the approximately constant stress of 100 dyne/cm^{-2} (with low viscosity products), or shear stress at an approximately constant shear rate of 10 s^{-1} (with high viscosity products).

Shama and Sherman (1973b) confirmed their earlier findings using distance and ratio scaling as the sensory response. Furthermore, they found that the change in oral stimulus from shear rate to shear stress begins at a viscosity of about 70 Hz.

The same authors have also shown that the stimulus controlling the sensory evaluation of viscosity may be different depending upon the sensory method of evaluation. As discussed above, oral evaluation may involve shear stress or shear rate. Non-oral evaluation by tilting the container was found to involve the shear rate (0.1–40 s^{-1}) developed at a shear stress (60–600 dyne /cm^{-2}) related to the flow properties of the sample. When judging viscosity by stirring the sample with a spoon, the stimulus was determined to be the shear stress developed at a particular rate of shear (90–100 s^{-1}) (Shama et al., 1973). Thus, with non-Newtonian fluids different methods of sensory evaluation may result in different ranking of test substances. In attempting correlations with objective measurements, not only should the appropriate test conditions be selected carefully, but they must reflect also the adopted sensory evaluation method.

11.3. 'ISOYPHS' METHOD

An unpublished report by Quarmby (1971) described a combination of a mathematical-graphical approach to the development of a relationship between subjective and objective measurements on potatoes. The report used the information in Table V to derive the following mathematical equation for the prediction of a sensory texture score for potatoes:

$$Y = -39.38 + 7.054X_1 - 0.2483X_2 - 0.2639X_1^2$$
$$- 0.000747X_2^2 + 0.02639X_1X_2, \qquad (9)$$

where Y = texture score, (0–10); X_1 = weighted mean, whole tuber total starch TS values (%); and X_2 = mean soluble starch, MSS, values (ppm).

TABLE V

Texture scores, total starch, and free soluble starch contents
(Quarmby, 1971)

Variety	Texture score (0–10)	Mean whole tuber TS (%)	MSS (ppm)
Kennebec	8.13	15.48	48.4
Exton	7.94	14.62	36.3
Kennebec D	7.50	14.90	75.3
Sebago	7.00	16.49	65.1
B550-1	6.69	11.89	32.0
9-1 (+K)	5.75	18.76	73.7
V59-4-8	5.63	12.79	47.6
V59-4-47	2.44	13.05	93.5
9-1 (-K)	1.38	21.59	140.9

The variables X_1 and X_2 represent two objective parameters as measured by the 'Sykes test', The value of the work at this stage lies in the approach to the problem of potato texture rather than in quantitative features of a final method; for the latter purpose much more confirmatory work is necessary. Figure 8 represents a plot of values of texture scores Y which result from using different X_1 and X_2 values in the above equation (note similarity to the response surface technique). The lines of con-

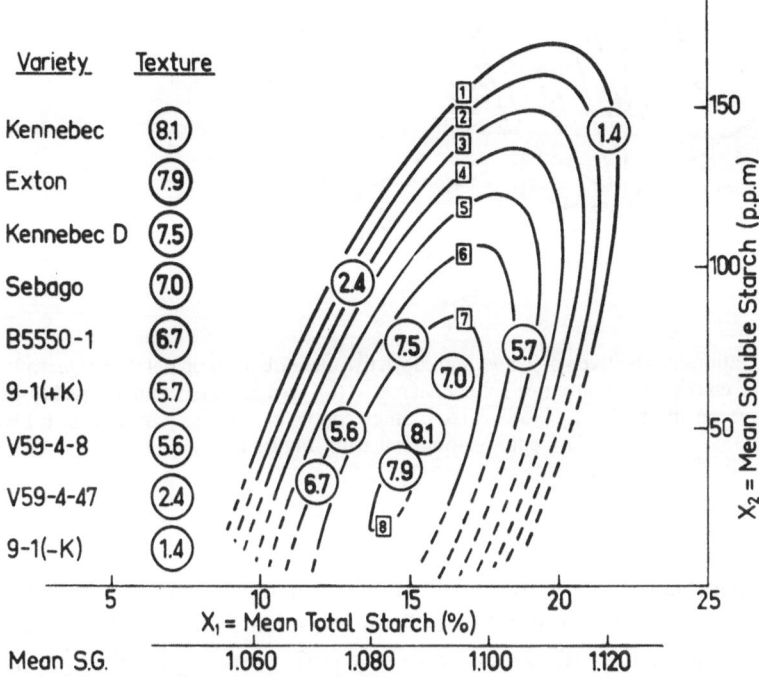

Fig. 8.   Texture 'isoyphs' of sensory texture scores on potatoes predicted from objective measurements
(Quarmby, 1971).

stant texture values ('isoyphs') give the levels of culinary quality predicted by the equation, with the circled numbers indicating the position of each variety with regard to its sensory texture score. It is evident that an increase in *MSS* values results in lower quality. An eight minute cooking time was used in all these cases. Figure 9a shows the effect of different cooking times on the *MSS* on one variety. When this curve is plotted in combination with the curves of Figure 8, the intercept of the cooking time curve gives a predicted texture score of approximately 5.0 for the Brownell tubers (Figure 9b). Since the eight-minute point in Figure 9a is close to the minimum of the curve, the movement of the plane of the isoyphs forward or backward (i.e. projection of the isoyphs toward either shorter or longer cooking times) will result in higher *MSS* values indicative of poorer quality.

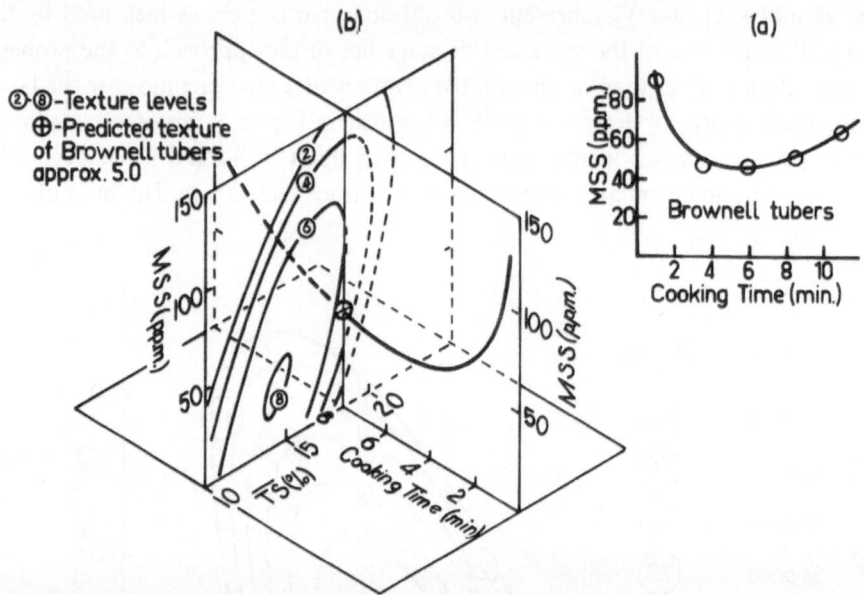

Fig. 9.   Relationships between objective and subjective measurements on potatoes (Quarmby, 1971). –
(a) Effect of cooking time on mean soluble starch. – (b) Prediction of culinary quality for a given
cooking time on the basis of objective tests. Interception of cooking curve from (a) with 8-min
cook orthograph from Figure 8.

11.4 REPRESENTATION IN SPACE

Three-dimensional representation of texture data is useful in the rapid assessment of
trends and relations which might be difficult to evaluate from tables or two-dimensional plots alone. Recently James and Olley (1971) used mechanical measurements
by an instrument (the Maturometer) to map the anatomy of muscle tissue of abalone
(a sea mollusk). In Figure 10 (A and B) the horizontal plaque represents a section of
muscle that was cut vertically from the central portion of the total mass. The heights
of the four 'cross-walls' represent Maturometer readings at the respective points of
the plaque. The variation of textural quality within the muscle section, and the in-
version of the texture profile between raw and canned abalone (Figure 10A and 10B)
can be clearly seen.

11.5. ANALYSIS OF SEVERAL ASSOCIATED DEPENDENT VARIABLES

In contrast to regression analysis which assumes a single dependent variable, there
may be studies involving two or more dependent variables, as in the case where more
than one distinct sensory (i.e. dependent) variable is considered simultaneously. The
statistical procedures are mathematically far more complex and the underlying assumptions are far more restrictive than for the regression model, in order to interrelate
the several dependent variables. However, it could have been expected that some
researchers will begin to experiment with multivariate statistics with the aid of com-

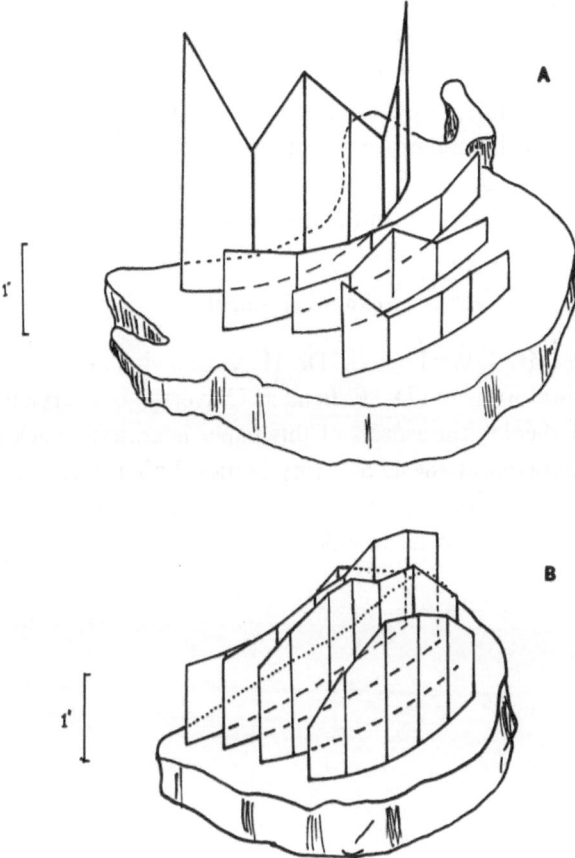

Fig. 10. Axonometric diagram showing texture profiles of abalone muscles (James and Olley, 1971). – (A) Raw Abalone. – (B) Canned Abalone.

puter programs in those instances where regression analysis yields unsatisfactory results. For a further discussion of multivariate statistics, see Kendall and Stuart (1966).

12. Conclusions

Methods for the study of the association between objective and subjective texture variables include: (a) plotting of the two variables in graphical form, (b) statistical treatment by means of correlation and regression analysis, and (c) mathematical treatment in cases where an exact mathematical relationship exists (probably in psychophysics laws). Statistical analysis makes possible the prediction of one variable from values of the other variable(s) within certain confidence limits at a specified level of probability. Due to fundamental differences between the mechanical property which a machine measures and the quality attribute which human senses perceive, the association between objective and subjective texture measurements is a 'covariation' caused by underlying common effects, rather than a direct relationship between two different types of measurement on the same property. There is need for more work on (a) correlation between rheological properties and sensory texture qualities, (b) rheological analysis of integrative or empirical texture instruments, and (c) psychophysical analysis of the human process of chewing. Parallel to advances in these three areas, consumer or product orientated texture instruments will continue to be useful to the food industry for development and quality control purposes.

Acknowledgements

The assistance of Mr. S. Werkowski, Dr. H. Moskowitz and Mr. R. Segars, U.S. Army Natick Laboratories, and Dr. F. Bender, University of Maryland, on statistical-mathematical and rheological aspects of this paper is gratefully acknowledged. Part of this paper undertaken at the U.S. Army Natick Laboratories has been assigned no. TP-1253 in the series of papers approved for publication. The findings in this report are not to be construed as an official Department of the Army position.

GENERAL BIBLIOGRAPHY

COMPILED BY MALCOLM C. BOURNE

Abbott, Judith A.: 1972, 'Sensory Assessment of Food Texture', *Food Technol.* **26**, 40.

Abbott, Judith A., Bachman, G. S., Childers, R. F., Fitzgerald, J. V., and Matusik, F. J.: 1968a, 'Sonic Techniques for Measuring Texture of Fruits and Vegetables', *Food Technol.* **22**, 635.

Abbott, Judith A., Childers, N. F., Bachman, G. S., Fitzgerald, J. V., and Matusik, F. J.: 1968b, 'Acoustic Vibration for Detecting Textural Quality of Apples', *Proc. Am. Soc. Hort. Sci.* **93**, 725.

Acton, F. S.: 1959, *Analysis of Straight-Line Data*, John Wiley & Sons, Inc. New York.

Adams, M. C. and Birdsall, E. L.: 1946, 'New Consistometer Measures Corn Consistency', *Food Inds.* **18**, 844.

Adams, Ruby, Harrison, Dorothy L., and Hall, J. L.: 1960, 'Comparison of Enzyme and Waring Blendor Methods for Determination of Collagen in Beef', *J. Agr. Food Chem.* **8**, 229.

Alfrey, T., Jr.: 1957, *Mechanical Behavior of High Polymers*, Interscience Publishers, Inc. New York, 581 pp.

American Association of Cereal Chemists: 1949, *Cereal Laboratory Methods*, p. 16.

American Society for Testing and Materials: 1967, 'Correlation of Subjective-Objective Methods in the Study of Odors and Taste', *ASTM Spec. Tech. Publ.*, No. 440.

American Society for Testing and Materials: 1968a, 'Manual on Sensory Testing Methods', *ASTM Spec. Tech. Publ.*, No. 434.

American Society for Testing and Materials: 1968b, 'Physical and Mechanical Testing of Metals; Nondestructive Tests', Part 31 of ASTM Standards, p. 150.

American Society for Testing and Materials: 1969, 'Reviews of Correlations of Objective-Subjective Methods in the Study of Odors and Taste', *ASTM Spec. Publ.*, No. 451.

Amerine, M. A., Pangborn, Rose Marie, and Roessler, E. B.: 1965, *Principles of Sensory Evaluation of Food*, Academic Press, New York, p. 505.

Anderson, D. J.: 1953, 'A Method of Recording Masticatory Loads', *J. Dent. Res.* **32**, 785.

Anderson, D. J. and Picton, D. C. G.: 1957, 'Tooth Contact During Chewing', *J. Dent. Res.* **36**, 21.

Anderson, P., Rapp, J., and Costello, D.: 1972a, 'Rotating Dull Knife Tenderometer', *Food Technol.* **26**, 25.

Anderson, P., Rapp, J., and Costello, D.: 1972b, 'The Problem of Devising a Total Tenderness Score', *J. Texture Studies* **3**, 122.

Ang, J. K., Isenberg, F. M., and Hartman, J. D.: 1960, 'Measurement of Firmness of Onion Bulbs with a Shear Press and a Potentio-Metric Recorder', *Proc. Am. Soc. Hort. Sci.* **75**, 500.

Angel, S. and Kramer, A.: 1969, 'Relation of Rheological Properties of *Pisum sativum* L. to Histological Changes', *J. Texture Studies* **1**, 90.

Angel, S., Kramer, A., and Yeatman, J. N.: 1965, 'Physical Methods of Measuring Quality of Canned Peas', *Food Technol.* **19**, 1278.

Anonymous: 1967, 'Meat-Textured Protein in Commercial Production', *Food Processing* **28**, 46 (February).

Anonymous: 1968, 'Measuring the Texture of a Swiss Roll', *New Scientist* **40**, 375 (November 14).

Anonymous: 1970, 'Tests Carcass Tenderness', *Food Eng.* **42**, 138.

Arbuckle, W. S.: 1960, 'The Microscopical Examination of the Texture and Structure of Ice Cream', *Ice Cream Trade J.* **56**, 62.

Askovitz, S. I.: 1957, 'A Short-Cut Graphic Method for Fitting the Best Straight Line to a Series of Points According to the Criterion of Least Squares', *J. Am. Statistical Assoc.* **52**, 13.

Association of Official Analytical Chemists: 1970, *Official Methods of Analysis*, 11th edition.

Avrami, M.: 1939, 'Kinetics of Phase Change. I: General Theory', *J. Chem. Phys.* **7**, 1103.

Avrami, M.: 1940, 'Kinetics of Phase Change. II: Transformation-Time Relations for Random Distribution of Nuclei', *J. Chem. Phys.* **8**, 212.

A. Kramer and A. S. Szczesniak (eds.), Texture Measurements of Foods, 161–175. All Rights Reserved.
Copyright © 1973 by D. Reidel Publishing Company, Dordrecht-Holland.

Avrami, M.: 1941, 'Kinetics of Phase Change. III: Granulation, Phase Change, Microstructure', *J. Chem. Phys.* **9**, 177.

Babb, A. T. S.: 1965, 'A Recording Instrument for the Rapid Evaluation of the Compressibility of Bakery Goods', *J. Sci. Fd. Agric.* **16**, 670.

Backinger, G. T., Kramer, A., Decker, R. W., and Sidwell, A. P.: 1957, 'Application of Work Measurement to the Determination of Fibrousness in Asparagus', *Food Technol.* **11**, 583.

Badgley, G. D.: 1967, 'Method and Apparatus for Measuring Tenderness', U.S. Patent 3 308 654.

Barton, R. R.: 1953, 'Colloids Help Improve the Quality of Frozen Raspberries', *Food Packer* **34**, 50.

Beebe-Center, J. G.: 1949, 'Standards for Use of the Gust Scale', *J. Psychology* **28**, 411.

Beebe-Center, J. G. and Waddell, D.: 1948, 'A General Psychological Scale of Taste', *J. Psychology* **26**, 517.

Better Homes & Gardens Encyclopedia of Cooking: 1970, **1**, 81.

Bewersdorff, H. J.: 1969, 'Elektronische dreidimensionale Messung und Registrierung von Kieferbewegungen', (Electronic Three Dimensional Measuring and Recording of Jaw Movements). Inaug. diss., Karol. Inst., Stockholm, Sweden, 94 pp.

Bice, C. W. and Geddes, W. F.: 1949, 'Studies in Bread Staling. IV: Evaluation of Methods for the Measurement of Changes which Occur During Bread Staling', *Cereal Chem.* **26**, 440.

Birth, G. S. and Norris, K. H.: 1958, 'An Instrument Using Light Transmittance for Nondestructive Measurement of Fruit Maturity', *Food Technol.* **12**, 592.

Bjorksten, J., Anderson, P., Bouschart, K. A., and Kapsalis, J.: 1967, 'A Portable Rotating Knife Tenderometer', *Food Technol.* **21**, 84.

Bland, D. R.: 1960, *The Theory of Linear Viscoelasticity*, Pergamon Press, New York, 125 pp.

Bloom, O. T.: 1925, 'Machine for Testing Jelly Strength of Glues, Gelatins, and the Like', U.S. Patent 1 540 979.

Bloom, O. J.: 1938, 'Consistency Tester', U.S. Patent 2 119 699.

Board, P. W., Gallop, R. A., and Sykes, S. M.: 1966, 'Quality of Canned Berry Fruits. I: The Influence of Sucrose Concentration and of Low Methoxyl Pectin Added to the Syrup', *Food Technol.* **20**, 1203.

Bockian, A. H., Anglemier, A. F., and Sather, Lois A.: 1958, 'Comparison of an Objective and Subjective Measurement of Beef Tenderness', *Food Technol.* **12**, 483.

Boggs, Mildred M., Campbell, H., and Schwartze, C. D.: 1943, 'Factors Influencing Texture of Peas Preserved by Freezing, II', *Food Res.* **8**, 502.

Boring, E. G.: 1942, *Sensation and Perception in the History of Experimental Psychology*, Appleton, Century Crofts, Inc.

Bourne, M. C.: 1965, 'Studies on Punch Testing of Apples', *Food Technol.* **19**, 413.

Bourne, M. C.: 1966a, 'A Classification of Objective Methods for Measuring Texture and Consistency of Foods', *J. Food Sci.* **31**, 1011.

Bourne, M. C.: 1966b, 'Measurement of Shear and Compression Components of Puncture Tests', *J. Food Sci.* **31**, 282.

Bourne, M. C.: 1967, 'Deformation Testing of Foods. I: A Precise Technique for Performing the Deformation Test', *J. Food Sci.* **32**, 601.

Bourne, M. C. and Moyer, J. C.: 1968, 'The Extrusion Principle in Texture Measurement of Fresh Peas', *Food Technol.* **22**, 1013.

Bouschart, K. and Meyer, E. A.: 1965, 'Apparatus for Testing Meat and the Like', U.S. Patent 3 214 967.

Box, G. E. P.: 1954, 'The Exploration and Exploitation of Response Surfaces. Some General Considerations and Examples', *Biometrics* **10**, 16.

Brabender, C. W.: 1965, 'Physical Dough Testing', *Cereal Science Today* **10**, 291.

Bradley, B. F.: 1966, 'Factors Influencing the Drained Weight, Texture and Other Processing Characteristics of Canned Strawberries,' *J. Sci. Food Agr.* **17**, 226.

Brandt, Margaret A., Skinner, Elaine Z., and Coleman, J. A.: 1963, 'Texture Profile Method', *J. Food Sci.* **28**, 404.

Bratzler, L. J.: 1932, 'Measuring the Tenderness of Meat by Means of a Mechanical Shear', M.S. Thesis, Kansas State University.

Brennan, J. G., Jowitt, R., and Mughsi, O. A.: 1970, 'Some Experiences with the General Foods Texturometer. An Interim Report', *J. Texture Studies* **1**, 167.

Briskey, E. J., Sayre, R. N., and Cassens, R. G.: 1962, 'Development and Application of an Apparatus for Continuous Measurement of Muscle Extensibility and Elasticity Before and During Rigor Mortis', *J. Food Sci.* **27**, 560.

Brody, A. L.: 1957, 'Masticatory Properties of Foods by Strain Gage Denture Tenderometer', Ph.D. Thesis, Mass. Institute Technology.

Buttkus, H. and Tarr, H. L. A.: 1962, 'Physical and Chemical Changes in Fish Muscle During Cold Storage', *Food Technol.* **16**, 84.

Caffyn, J. E. and Baron, Margaret: 1947, 'Scientific Control in Cheese Making', *The Dairyman* **64**, 345.

Cairncross, S. E. and Sjöström, L. B.: 1950, 'Flavor Profiles – A New Approach to Flavor Problems', *Food Technol.* **4**, 308.

Caldwell, J. S.: 1939, 'Factors Influencing Quality of Sweet Corn', *Canning Trade* **61**, 7.

Casimir, D. J., Coote, G. G., and Moyer, J. C.: 1971, 'Pea Texture Studies Using a Single Puncture Maturometer', *J. Texture Studies* **2**, 419.

Chappell, T. W. and Hamann, D. D.: 1968, 'Poisson's Ratio and Young's Modulus for Apple Flesh Under Compressive Loading', *Trans. Am. Soc. Ag. Eng.* **11**, 608.

Charm, S. E.: 1960, 'Viscometry of Non-Newtonian Food Materials', *Food Res.* **25**, 351.

Charm, S. E.: 1962, 'The Nature and Role of Fluid Consistency in Food Engineering Applications', Academic Press, New York, p. 355.

Charm, S. E.: 1963, 'The Direct Determination of Shear Stress – Shear Rate Behavior of Foods in the Presence of a Yield Stress', *J. Food Sci.* **28**, 107.

Charm, S. E.: 1964, 'The Determination of the Tensile Strength of Fluid Food Materials', *J. Food Sci.* **29**, 483.

Charm, S. E. and McComis, W.: 1965, 'Physical Measurements of Gums', *Food Technol.* **19**, 948.

Chopin, M. J. E.: 1936, 'Procedure and Instrument for Measuring Flour Pastes and Other Plastic Substances', French Patent 799966.

Chopin, M. J. E.: 1965, 'Procedure and Instrument for Measuring Flour Pastes and Other Plastic Substances; French Patent 1397495.

Christel, W. F.: 1938, 'Texturemeter, a New Device for Measuring the Texture of Peas', *Canning Trade* **60**, 10.

Clark, J. H.: 1928, *49th Ann. Rep. N. J. State Agr. Exp. Sta.* 227.

Cochran, W. G. and Cox, Gertrude M.: 1957, *Experimental Designs*, John Wiley & Sons, Inc., New York.

Coombs, C. H.: 1952, 'A Theory of Psychological Scaling', *Eng. Res. Bull.* **34**, University of Michigan Press, Ann Arbor.

Coombs, C. H.: 1964, *A Theory of Data*, John Wiley & Sons, Inc., New York.

Cooper, H. E.: 1962, 'Influence of Maturation on the Physical and Mechanical Properties of Apple Fruit', M.S. Thesis, The Pennsylvania State University.

Corey, H.: 1970a, 'Texture in Foodstuffs', *Crit. Rev. Food Technol.* **1**, 161. Chemical Rubber Co., Cleveland.

Corey, H.: 1970b, 'On the Texture of Foodstuffs', Proc. 3rd Int'l Congress Food Sci. and Technol. Washington, D.C. Aug. 9–14.

Corey, H. and Creswick, N.: 1970, 'A Versatile Recording Couette-Type Viscometer', *J. Texture Studies* **1**, 155.

Cornford, S. J., Axford, D. W. E., and Elton, G. A. H.: 1964, 'The Elastic Modulus of Bread Crumb in Linear Compression in Relation to Staling', *Cereal Chem.* **41**, 216.

Cover, Sylvia, Ritchey, S. J., and Hostetler, R. L.: 1962, 'Tenderness of Beef. Parts I–IV', *J. Food Sci.* **27**, 469, 527.

Cowie, W. P. and Little, W. T.: 1966, 'The Relationship Between Toughness of Cod Stored at − 29 °C and Its Muscle Protein Solubility and pH', *J. Food Technol.* **1**, 335.

Cox, C. P. and Baron, Margaret: 1955, 'A Variability Study of Firmness in Cheese Using the Ball Compressor Test', *J. Dairy Res.* **22**, 386.

Crocker, E. C.: 1945, *Flavor*, McGraw Hill Book Co., Inc., New York, 172 pp.

Dassow, J. A., McKee, Lynne G., and Nelson, R. W.: 1962, 'Development of an Instrument for Texture Evaluation of Fishery Products', *Food Technol.* **3**, 108.

Davey, C. L. and Gilbert, K. V.: 1969, 'The Effect of Sample Dimensions on the Cleaving of Meat in the Objective Assessment of Tenderness', *J. Food Technol.* **4**, 7.

Davies, O. L.: 1954, *The Design and Analysis of Industrial Experiments*, Oliver and Boyd, Ltd. Edinburgh.

Davis, C. E.: 1921, 'Shortening: Its Definition and Measurement', *Ind. Eng. Chem.* **13**, 797.

Davis, R. B., De Weese, D., and Gould, W. A.: 1954, 'Consistency Measurements of Tomato Puree', *Food Technol.* **8**, 330.

Davis, W. B.: 1969, 'Fruit Ripeness Determination', U.S. Patent 3420635.

Dawson, Elsie H. and Harris, Betsy L.: 1951, 'Sensory Methods for Measuring Differences in Food Quality', USDA Ag. Info. Bul. No. 34.

DeFremery, D. and Pool, M. F.: 1960, 'Biochemistry of Chicken Muscle as Related to Rigor Mortis and Tenderization', *Food Res.* **25**, 73.

de Man, J.: 1969, 'Food Texture Measurements with the Penetration Method', *J. Texture Studies* **1**, 114.

Drake, B. K.: 1961, 'An Attempt at a Geometrical Classification of Rheological Apparatus', unpublished ms. (quoted by Bourne, 1966a).

Drake, B. K.: 1962, 'Automatic Recording of Vibrational Properties of Foodstuffs', *J. Food Sci.* **27**, 182.

Drake, B. K.: 1965a, 'On the Biorheology of Human Mastication: An Amplitude-Frequency-Time Analysis of Food Crushing Sounds', *Biorheol.* **3**, 21.

Drake, B. K.: 1965b, 'Food Crushing Sounds: Comparions of Objective and Subjective Data', *J. Food Sci.* **30**, 556.

Drake, B. K.: 1966, 'Advances in the Determination of Texture and Consistency of Foodstuffs', Proc. 2nd Int'l Congress Food Sci. and Technol., Warsaw, Poland, p. 277.

Drake, B. K.: 1968, 'The Biorheological Process of Mastication'. Rheology and Texture of Foodstuffs, Monograph No. 27. Soc. Chem. Ind. London, p. 29.

Dunlop, C. A. A. and Ormrod, D. P.: 1970, 'Temperature Effects During Fruit Development on the Quality of Green Snap Beans', *J. Can. Inst. Food Technol.* **3**, 6.

Dwyer, P. S.: 1951, *Linear Computations*, John Wiley & Sons, Inc., New York.

Efroymsen, M. A.: 1960, 'Multiple Regression Analysis', in *Mathematical Methods for Digital Computers* (ed. by A. Ralston and H. S. Wilf), John Wiley & Sons, New York, p. 191.

Eirich, F. R. (ed.): 1956, *Rheology – Theory and Applications*, Vols. I–III, Academic Press, New York.

Ekman, G., Eisler, H., and Künnapas, T.: 1960, 'Brightness Scales for Monochromatic Light', *Scand. J. Psychology* **1**, 41.

Ekman, G., Hosman, J., and Lindström, B.: 1964, 'Roughness, Smoothness and Preference: a Study of Quantitative Relations in Individual Subjects', *J. Exp. Psychology* **70**, 18.

Elder, A. L. and Smith, R. J.: 1969, 'Food Rheology Today', *Food Technol.* **23**, 629.

Ellis, B. H.: 1966, *Guide Book for Sensory Testing*, Continental Can Company Manual.

Ellis, B. H.: 1967, 'Efficient Use of Sensory Evaluation Methods', *Food Prod. Develop.* Oct.–Nov.

El-Sayed, M. N. K., Erickson, H. T., and Tomes, M. L.: 1966, 'Pectic Substances in Tomatoes as Related to Whole-Fruit Firmness and Inheritance', *Proc. Am. Soc. Hort. Sci.* **89**, 528.

Elton, G. A. H.: 1969, 'Some Quantitative Aspects of Bread Staling, *Bakers Digest.* **43**, 24, 76.

Eolkin, D.: 1957, 'The Plastometer – a New Development in Continuous Recording and Controlling Consistometers', *Food Technol.* **11**, 253.

Ernest, Jane V., Birth, G. S., Sidwell, A. P., and Golumbic, C.: 1958, 'Evaluation of Light Transmittance Technique for Maturity Measurements of Two Varieties of Prune-Type Plums. *Food Technol.* **12**, 595.

Etienne, J. J. and Dubois, M.: 1970, 'Studies of Paste Consistency with the Aid of Chopin Teleplastometer', *Bull. Anciens Eleves Ecole Fr. Meun No. 236*, 94.

Ezekiel, M. and Fox, K. A.: 1959, *Methods of Correlation and Regression Analysis, Linear and Curvilinear*, 3rd ed., John Wiley & Sons, Inc., New York.

Ezell, G. H.: 1959, 'Viscosity of Concentrated Orange and Grape-Fruit Juice', *Food Technol.* **13**, 9.

Falk, S., Hertz, C. H., and Virgin, H. I.: 1958, 'On the Relation Between Turgor Pressure and Tissue Rigidity', *Physiologia Plantarum* **11**, 802.

Farkas, Elizabeth and Glicksman, M.: 1967, 'Hydrocolloid Rheology in the Formulation of Convenience Foods', *Food Technol.* **21**, 535.

Ferry, J. D.: 1961, *Viscoelastic Properties of Polymers*, John Wiley & Sons, New York, p. 105.

Fielder, Mary M., Mullins, A. M., Skellenger, Marie M., Whitehead, Ruby, and Moschette, Dorothy S.: 1963, 'Subjective and Objective Evaluations of Prefabricated Cuts of Beef', *Food Technol.* **17**, 213.

Finney, E. E., Jr.: 1967, 'Dynamic Elastic Properties of Some Fruits During Growth and Development', *J. Agr. Eng. Res.* **12**, 249.

Finney, E. E., Jr.: 1969a, 'Objective Measurements for Texture in Foods,' *J. Texture Studies* **1**, 19.

Finney, E. E., Jr.: 1969b, 'To Define Texture in Fruits and Vegetables', *Ag. Eng.* **50**, 462.

Finney, E. E., Jr.: 1971, 'Dynamic Elastic Properties and Sensory Quality of Apple Fruit', *J. Texture Studies* **2**, 62.

Finney, E. E., Jr. and Hall, C. W.: 1964, 'Der Einfluss der Belastungsfläche bei mechanischer Zerstörung des Kartoffelgewebes', *Landtechnische Forsch.* **14**, 161.

Finney, E. E., Jr. and Hall, C. W.: 1967, 'Elastic Properties of Potatoes', *Trans. Am. Soc. Ag. Eng.* **10**, 4.

Finney, E. E., Jr. and Norris, K. H.: 1967, 'Sonic Resonant Methods for Measuring Properties Associated with Texture of Irish and Sweet Potatoes', *Proc. Am. Soc. Hort. Sci.* **90**, 275.

Finney, E. E., Jr. and Norris, K. H.: 1968, 'Instrumentation for Investigating Dynamic Mechanical Properties of Fruits and Vegetables', *Trans. Am. Soc. Agr. Eng.* **11**, 94.

Finney, E. E., Jr., Hall, C. W., and Mase, G. E.: 1964, 'Theory of Linear Viscoelasticity Applied to the Potato', *J. Agr. Eng. Res.* **9**, 307.

Fisher, R. A.: 1970, *Statistical Methods for Research Workers*, 14th ed., Hafner Publishing Co., Darien, Conn.

Fitzgerald, E. R.: 1961, 'Yield Strength of Crystalline Solids from Dynamic Mechanical Measurements', *Development in Mechanics* **1**, 10 (ed. by J. E. Lay and L. E. Malvern), Plenum Press, New York, 611 pp.

Fortini, Bonnie and Hogan, J. M.: 1966, 'Shear Tests as a Measure of Mealiness of Potatoes', *Maine Farm Res.* **14**, 15.

Fox, M. and Kramer, A.: 1966, 'Objective Tests for Determining Quality of Fresh Green Beans', *Food Technol.* **20**, 88.

Friedman, H. H., Whitney, J. E., and Szczesniak, Alina S.: 1963, 'The Texturometer – a New Instrument for Objective Texture Measurement', *J. Food Sci.* **28**, 390.

Funk, Kaye, Zabik, Mary E., and Downs, Doris M.: 1965, 'Comparison of Shear-Press Measurements and Sensory Evaluation of Angel-Cakes', *J. Food Sci.* **30**, 729.

Funk, Kaye, Zabik, Mary E., and Elgidaily, Doha A.: 1969, 'Objective Measurements for Baked Products', *J. Home Economics* **61**, 119.

Gacula, M. C., Jr., Reaume, June B., Morgan, K. J., and Luckett, R. L.: 1971, 'Statistical Aspects of the Correlation Between Objective and Subjective Measurements of Meat Tenderness', *J. Food Sci.* **36**, 185.

Gairns, F. W.: 1951, 'Nerve Endings in the Human Gum and Hard Palate', *J. Physiol. (Lond.)* **115**, 70P.

Gairns, F. W.: 1954, 'Sensory Nerve Endings in the Human Palate', *J. Physiol. (Lond.)* **123**, 26P.

Garrett, A. W., Desrosier, N. W., Kuhn, G. D., and Fields, M. L.: 1960, 'Evaluation of Instruments to Measure Firmness of Tomatoes', *Food Technol.* **14**, 562.

General Foods Corporation: 1970, 'Sensory Texture Analysis', General Foods Corp. Manual.

Gordon, A.: 1967, 'Measurement of Texture in Heterogeneous Type Foods', Brit. Food Manuf. Ind. Res. Assoc. Sci. and Tech. Survey, No. 50.

Gordon, A.: 1969, 'Food Texture Survey of Some of the More Recent Work in This Field', *Food Proc. Marketing* **38**, 54.

Gould, W. A., Krantz, F. A., Jr., and Mavis, J.: 1951, 'Quality Evaluation of Fresh, Frozen and Canned Yellow Sweet Corn', *Food Technol.* **5**, 175.

Grau, R. and Hamm, R.: 1953, 'Determination of the Water Binding Power of Muscle', *Naturwissenschaften* **40**, 29.

Gross, Bernhard: 1953, *Mathematical Structure of the Theories of Viscoelasticity*, Hermann and Cie., Paris, 74 pp.

Grünewald, T.: 1957, 'Ein Festigkeitsprüfgerät für Lebensmittel nach N. Wolodkewitsch', *Z. Lebensm. Untersuch. Forsch.* **105**, 1.

Guilford, J. P.: 1965, *Fundamental Statistics in Psychology and Education*, 4th ed., Chapter 14, McGraw-Hill Publishing Co., New York.

Gutschmidt, J.: 1954, 'A Contribution to the Determination of Maturity Grade of Green Peas by Means of the Texturemeter. III: Comparative Determinations of the Degree of Maturity', *Industr. Obst. v. Gemuseverw.* **39**, 242.

Haighton, A. J.: 1959, 'The Measurement of the Hardness of Margarine and Fats with Cone Penetrometers', *J. Amer. Oil Chem. Soc.* **36**, 345.

Hall, R. C. and Fryer, H. C.: 1953, 'Consistency Evaluation of Dehydrated Potato Granules and Direction for Microscopic Rupture Count Procedure', *Food Technol.* **7**, 373.

Haller, M. H.: 1941, 'Fruit Pressure Testers and Their Practical Applications', USDA Circ. 627.

Halton, P.: 1949, 'Significance of Load-Extension Tests in Assessing the Baking Quality of Wheat Flour Doughs', *Cereal Chem.* **26**, 24.

Halton, P. and Scott Blair, G. W.: 1937, 'A Study of Some Physical Properties of Flour Doughs in Relation to Their Bread-Making Properties', *Cereal Chem.* **14**, 201.

Hamann, D. D.: 1969, 'Dynamic Mechanical Properties of Apple Fruit Flesh', *Trans. Am. Soc. Ag. Eng.* **12**, 170.

Hamm, R. and Kulmach, K. R.: 1967, *The Rheology of Meat. I: The Measurement of Flow Propertief in Muscle Homogenates Using the Rotation Viscosimeter*, (translated from *Die Fleisch Wirtschast* by R. G. Menzel), p. 1122.

Hammerle, J. R.: 1968, 'Evaluating Egg Shell Strength Through Experimental Stress Analysis', *Trans. Am. Soc. Agr. Eng.* **12**, 703.

Hammerle, J. R. and McClure, W. F.: 1971, 'The Determination of Poisson's Ratio by Compression Tests of Cylindrical Specimens', *J. Texture Studies* **2**, 31.

Hansen, L. J.: 1972, 'Development of the Armour Tenderometer for Tenderness Evaluation of Beef Carcasses', *J. Texture Studies* **3**, 146.

Harman, H. H.: 1967, *Modern Factor Analysis*, The University of Chicago Press, 2nd ed.

Harper, R.: 1956, 'Factor Analysis as a Technique for Examining Complex Data on Foodstuffs', *Appl. Statistics* **5**, 32.

Harper, R.: 1968, 'Texture and Consistency from the Standpoint of Perception: Some Major Issues', in *Rheology and Texture of Foodstuffs*, Monograph No. 27, Soc. Chem. Ind. London, p. 11.

Harper, R. and Baron, Margaret: 1948, 'Factorial Analysis of Rheological Measurements on Cheese', *Nature* **162**, 821.

Harper, R. and Baron Margaret: 1951, 'The Application of Factor Analysis to Tests on Cheese', *Brit. J. Appl. Phys.* **2**, 35.

Harper, R. and Stevens, S. S.: 1964, 'Subjective Hardness of Compliant Materials', *Quart. J. Exper. Psychology.* **16**, 204.

Harrington, B. S.: 1948, 'Consistometer', U.S. Patent 2437194.

Heiss, R. and Witzel, H.: 1969, 'Objective Methods for Measurement of Consistency in Solid Foods', (in German) *Z. Lebensmittel-Unters. Forsch.* **141**, 87.

Henry, W. F., Katz, M. H., Pilgrim, F. J., and May, A. T.: 1971, 'Texture of Semi-Solid Foods: Sensory and Physical Correlates', *J. Food. Sci.* **36**, 155.

Herring, H. K., Cassens, R. G., and Briskey, E. J.: 1965, 'Sarcomere Length of Free and Restrained Bovine Muscles at Low Temperatures as Related to Tenderness', *J. Sci. Food Agr.* **16**, 379.

Hibberd, G. E. and Wallace, W. J.: 1966, 'Dynamic Viscoelastic Behavior of Wheat Flour Doughs. Part I. Linear Aspects', *Rheologica Acta* **5**, 193.

Hibberd, G. E., Wallace, W. J., and Wyatt, K. A.: 1966, 'A Rheometer for Measuring the Dynamic Mechanical Properties of Soft Solids', *J. Sci. Instr.* **43**, 84.

Hill, Rodney: 1960, *The Mathematical Theory of Plasticity*, Oxford University Press, New York, 355 pp.

Hintzer, H. M. R.: 1949, 'The Staling of Bread', *Chem. Weekbl.* **45**, 797.

Hoeppler, F.: 1954, 'Rheologische Messgeräte', *Forschungsberichte, VEB, Prüfgeräte-werk Medingen* No. 3.

Holt, C. B.: 1970, 'Measurement of Tomato Firmness with a Universal Testing Machine', *J. Texture Studies* **1**, 491.

Hopkins, J. W.: 1950, 'A Procedure for Quantifying Subjective Appraisals of Odor, Flavor and Texture of Foodstuffs', *Biometrics* **6**, 1.

Hostetler, R. L. and Cover, Sylvia: 1961, 'Relationship of Extensibility of Muscle Fibers to Tenderness of Beef', *J. Food Sci.* **26**, 535.

Howard, P. L. and Heinz, D. E.: 1970, 'Texture of Carrots', *J. Texture Studies* **1**, 185.

Huebner, V. R. and Thomsen, L. C.: 1957, 'Spreadability and Hardness of Butter. I: Development of an Instrument for Measuring Spreadability', *J. Dairy Sci.* **40**, 834.

Huff, E. R.: 1967, 'Tensile Properties of Kennebec Potatoes', *Trans. Am. Soc. Agr. Eng.* **10**, 414.

Isherwood, F. A.: 1955, 'Texture in Fruit and Vegetables', *Food Manuf.* **30**, 399.

James, D. G. and Olley, June: 1971, 'Studies on the Processing of Abalone. II: The Maturometer as a Guide to Canned Abalone Texture', *Food Technol. Aust.* **23**, 394.

James, P. E. and Retzer, H. J.: 1967, 'Measuring Egg Shell Strength by Beta Scatter Technique', *Poultry Sci.* **46**, 1200.

Jankelson, B., Hoffman, G. M., and Henderson, J. A.: 1953, 'The Physiology of the Stomatognathic System', *J. Amer. Dent. Assoc.* **46**, 378.

Jastrzebski, Z. D.: 1959, *Nature and Properties of Engineering Materials*, John Wiley & Sons, New York, p. 159.

Jellinek, Gisela: 1964, 'Introduction to and Critical Review of Modern Methods of Sensory Analysis (Odour, Taste, and Flavour Evaluation) with Special Emphasis on Descriptive Sensory Analysis (Flavor Profile Method)', *J. Nutrition and Dietetics* (India) **1**, 219.

Jellinek, H. H. G. and Brill, R.: 1956, 'Viscoelastic Properties of Ice', *J. Appl. Phys.* **27**, 1198.

Jenkins, G. N.: 1966, *The Physiology of the Mouth*, 3rd ed., Blackwell Scientific Publ., London.

Johansen, D. A.: 1940, *Plant Microtechnique*, McGraw-Hill, New York.

Jones, N. R.: 1968, 'Texture of Agar Jellies; Measurement and Modification', in *Rheology and Texture of Foodstuffs*, Soc. Chem. Ind. London, Monograph No. 27, p. 91.

Joslyn, M. A.: 1970, *Methods in Food Analysis*, 2nd ed., Academic Press, New York.

Kaldy, M. S., Bergen, H. J., and Ramage, Christina T.: 1967, 'A Washing Apparatus for Use in Fiber Determination', *Can. J. Plant Sci.* **47**, 335.

Kapsalis, J. G.: 1967, 'Hygroscopic Equilibrium and Texture of Freeze-Dried Foods. Investigation of the Relationships Between Moisture Content – Water Vapor Equilibrium and Textural Parameters in Special Freeze-Dried Foods', Defense Documentation Center, Tech. Report AD 655 488, Clearing house for Federal Scientific and Technical Information, U.S. Dept. of Commerce, Springfield, Va.

Kapsalis, J. G., Betscher, J. J., Kristoffersen, T., and Gould, I. A.: 1960, 'Effect of Chemical Additives on the Spreading Quality of Butter. I: The Consistency of Butter as Determined by Mechanical and Consumer Panel Evaluation Methods', *J. Dairy Sci.* **43**, 1560.

Kapsalis, J. G., Drake, B. K., and Johansson, B.: 1970, 'Textural Properties of Dehydrated Foods. Relationships with the Thermodynamics of Water Vapor Sorption', *J. Texture Studies* **1**, 285.

Kapsalis, J. G., Kramer, A., and Szczesniak, A. S.: 1973, 'Quantification of Objective and Sensory Texture Relations', this volume, p. 130.

Kapsalis, J. G., Kristoffersen, T., Gould, I. A., and Betscher, J. J.: 1963, 'Effect of Chemical Additives on the Spreading Quality of Butter. II: Laboratory and Plant Churnings', *J. Dairy Sci.* **46**, 107.

Kapsalis, J. G., Segars, R. A., and Krizik, Jan G.: 1972, 'An Instrument for Measuring Rheological Properties by Bending. Application to Materials of Plant Origin', *J. Texture Studies* **3**, 31.

Kapur, K. K.: 1971, 'Frequency Spectrographic Analysis of Bone Conducted Chewing Sounds in Persons with Natural and Artificial Dentitions', *J. Texture Studies* **2**, 50.

Kapur, K. K., Chauncey, H. H., and Sharon, I. M.: 1966, 'Oral Physiological Factors Concerned with Ingestion of Foods', in *The Science of Nutrition and Its Application in Clinical Dentistry*, (ed. by A. E. Nizell), 2nd ed., W. B. Saunders Co., Philadelphia.

Kastner, C. L. and Henrickson, R. L.: 1970, 'Providing Uniform Meat Cores for Mechanical Shear Force Measurement', *J. Food Sci.* **34**, 603.

Kattan, A. A.: 1957, 'Changes in Colour and Firmness During Ripening of Detached Tomatoes and the Use of a New Instrument for Measuring Firmness', *Proc. Am. Soc. Hort. Sci.* **70**, 379.

Kawamura, Y.: 1964, 'Recent Concepts of the Physiology of Mastication', *Adv. Oral. Biol.* **1**, 77.

Kelly, R. F., Taylor, J. C., and Graham, P. P.: 1960, 'Preliminary Comparisons of a New Tenderness Measuring Device with Objective and Subjective Evaluations of Beef', *J. Animal Sci.* **19**, 645 (abstract).

Kendall, M. G. and Stuart, A.: 1966, *Advanced Theory of Statistics*, Vol. III, Hafner Publ. Co., New York.

Kertesz, Z. I.: 1935, 'The Chemical Determination of the Quality of Canned Green Peas', *N. Y. State Agr. Expt. Sta. Tech. Bull.* 223.

Kissel, L. T.: 1967, 'Optimization of White Layer Cake Formulations by a Multiple-Factor Experimental Design', *Cereal Chem.* **44**, 253.

Koprowski, W. S.: 1951, 'Determination of Jelly Strength of Glues and Gelatins by the "Boucher" Jelly Tester', *Analyst* **76**, 732.

Kramer, A.: 1951, 'Faster Quality Check for Snap Beans', *Food Packer* **32**, 32.

Kramer, A.: 1955, 'Food Quality and Quality Control', Chapter 23, *Handbook of Food and Agriculture* (ed. by Blanck), Reinhold Publ. Co., New York, p. 733.

Kramer, A.: 1956, 'The Problem of Developing Grades and Standards of Quality', *Food, Drug, Cosmetic Law J.* **7**, 23.

Kramer, A.: 1961, 'The Shear-Press, a Basic Tool for the Food Technologist', *The Food Scientist* **5**, 7.

Kramer, A.: 1964, 'Definition of Texture and Its Measurement in Vegetable Products', *Food Technol.* **18**, 304.

Kramer, A.: 1965a, 'The Effective Use of Operations Research and EVOP in Quality Control', *Food Technol.* **19**, 37.

Kramer, A.: 1965b, 'Evaluation of Quality of Fruits and Vegetables. Food Quality: Effects of Production Practices and Processing', Publication 77, *Am. Assoc. Adv. Science*, p. 9.

Kramer, A.: 1966, 'Sensory Evaluation of Food Flavor. Flavor Chemistry', *Amer. Chem. Soc. Adv. in Chemistry Series 56*, Washington, D.C., p. 64.

Kramer, A.: 1968, 'The Judging of Food Quality – A Consideration of Uniform Scoring', *Proc. Tech. Mtg. Food and Dairy Ind. Expo.*, p. 79.

Kramer, A.: 1969, 'The Relevance of Correlating Objective and Subjective Data', *Food Technol.* **23**, 66.

Kramer, A.: 1973, 'Food Texture – Definition, Measurement and Relation to Other Food Quality Attributes', this volume, p. 1.

Kramer, A. and Hart, W. J., Jr.: 1954, 'Recommendations on Procedures for Determining Grades of Raw, Canned, and Frozen Lima Beans', *Food Technol.* **8**, 55.

Kramer, A. and Hawbecker, J. V.: 1966, 'Measuring and Recording Rheological Properties of Gels', *Food Technol.* **20**, 209.

Kramer, A. and Smith, H. R.: 1946, 'The Succulometer, an Instrument for Measuring Maturity of Raw and Canned Kernel Corn', *Food Packer* **27**, 56.

Kramer, A. and Twigg, B. A.: 1959, 'Principles and Instrumentation for the Physical Measurements of Food Quality with Special Reference to Fruit and Vegetable Products', *Adv. Food Research* **9**, 153.

Kramer, A. and Twigg, B. A.: 1970, *Fundamentals of Quality Control for the Food Industry*, 2nd ed., Avi Publ. Co., Westport, Conn., ch. 4 and 7.

Kramer, A., Haut, I. C., Scott, L. E., and Ide, L. E.: 1949, 'Quick Recorders of Fibrousness in Asparagus', *Food Inds.* **21**, 1075.

Kruskal, J. B.: 1964, 'Non-Metric Multidimensional Scaling: A Numerical Method', *Psykometrika* **29**, 115.

Kuhn, G., Desrosier, N. W., and Ammerman, G.: 1959, 'Relation of Chemical Composition and Some Physical Properties to Potato Texture', *Food Technol.* **13**, 183.

Kulwich, R., Decker, R. W., and Alsmeyer, R. H.: 1963, 'Use of a Slice-Tenderness Evaluation Device with Pork', *Food Technol.* **17**, 201.

LaBelle, R. L.: 1964, 'Bulk Density: a Versatile Measure of Food Texture and Bulk', *Food Technol.* **18**, 89.

LaBelle, R. L. and Moyer, J. C.: 1960, 'Factors Affecting the Drained Weight and Firmness of Red Tart Cherries', *Food Technol.* **14**, 347.

Lanza, J. and Kramer, A.: 1967, 'Objective Measurement of Graininess of Apple Sauce', *Proc. Am. Soc. Hort. Sci.* **90**, 491.

Larmond, E.: 1970, 'Methods for Sensory Evaluations of Food', Canada Dept. Ag. Pub. 1284.

Lawrie, R. A.: 1968, 'Factors Modifying Meat Texture', in *Rheology and Texture of Foodstuffs*, Soc. Chem. Ind. Monograph No. 27, London, p. 134.

Ledley, R. S.: 1971, 'Dental Forces and Mastication', *J. Texture Studies* **2**, 3.

Lee, F. A.: 1940, 'Determination of the Quality of Vegetables', *Proc. 1st Food Conf.*, Inst. Food Technologists, p. 33.

Lehmann, K. B.: 1907, 'Studies on the Toughness of Meat and Its Origin' (in German), *Arch. Hyg.* **63**, 134.

LeMagnen, J.: 1962, 'Vocabulaire technique des caractères organoleptiques et de la dégustation des produits alimentaires', *Annales de la Nutrition et de l'Alimentation* **18**, B1.

Leonard, S. J., Luh, B. S., and Mrak, E. M.: 1958, 'Factors Influencing Drained Weight of Canned Clingstone Peaches', *Food Technol.* **12**, 80.

Lerchenthal, C. H. and Muller, H. G.: 1967, 'Research in Dough Rheology at the Israel Institute of Technology', *Cereal Sci. Today* **12**, 185.

Linehan, D. J. and Hughes, J. C.: 1969, 'Texture of Cooked Potato. II: Relationships Between Intercellular Adhesion and Chemical Composition of the Tuber', *J. Sci. Food Agr.* **20**, 113.

Lingoes, J. C.: 1966, 'An IBM 7090 Program for Guttman-Lingoes Smallest Space Analysis III', *Behavioral Science* **11**, 75.

Litchfield, J. T., Jr. and Wilcoxon, F.: 1955, 'The Rank Correlation Method', *Anal. Chem.* **27**, 299.

Lockwood, H. C. and Hayes, R. S.: 1931, 'A New Method for Testing Agar and Gelatin Jellies', *J. Soc. Chem. Ind.* **50**, 145.

Love, R. M.: 1960, 'Texture Change in Fish and Its Measurement', in *Texture in Foods*, Soc. Chem. Ind. Monograph No. 7, London, p. 109.

Lovegren, N. V., Guice, W. A., and Feuge, R. O.: 1958, 'An Instrument for Measuring the Hardness of Fats and Waxes', *J. Am. Oil Chemists' Soc.* **35**, 327.

Lundgren, H. P.: 1969, 'New Concepts on Evaluating Fabric', *Hand. Textile Chemist Colorist* **1**, 35.

Lundstedt, E.: 1955, 'New Curd Meter Takes Guesswork out of Cheese Making', *Food Eng.* **27**, 97.

MacAllister, R. V. and Reichenwallner, C.: 1959, 'Force-Deformation Measuring Apparatus', U.S. Patent 2912855.

MacDonald, I. A. and Bly, D. A.: 1966, 'Determination of Optimal Levels of Several Emulsifiers in Cake Mix Shortenings', *Cereal Chem.* **43**, 571.

MacMichael, R. F.: 1915, 'A New Direct-Reading Viscosimeter', *Ind. Eng. Chem.* **7**, 961.

Magness, J. R. and Taylor, G. F.: 1925, 'An Improved Type of Pressure Tester for the Determination of Fruit Maturity', U.S. Dept. Agr. Circ. 350.

Mahon, J. H. and Schneider, C. G.: 1964, 'Minimizing Freezing Damage and Thawing Drip on Fish Fillets', *Food Technol.* **18**, 1941.

Malvern, L. E.: 1962, 'Constitutive Equations of Elasticity, the Elastic Potentials or Strain Energy Function, and Generalized Hook's Law', Unpublished class notes for *Applied Mechanics* 813 – Elasticity, Michigan State University, 9 pp.

Martin, W. McK.: 1937, 'The Tenderometer, an Apparatus for Evaluating Tenderness in Peas', *Canning Trade* **59**, 29 (February 22).

Martin, W. McK.: 1941, 'Device for Determining the Consistency of Food Mixtures, Paints, or Other Materials', U.S. Patent 2239726.

Matthews, Ruth H. and Dawson, Elsie H.: 1963, 'Performance of Fats and Oils in Pastry and Biscuits', *Cereal Chem.* **40**, 291.

Matz, S. A.: 1962, *Food Texture*, Avi Pub. Co., Westport, Conn., 286 pp.

Miller, B. S. and Trimbo, H. B.: 1970, 'Factors Affecting the Quality of Pie Dough and Pie Crust', *Bakers Digest* **44**, 46, 52.

Milleville, H. P. and Leinen, N. J.: 1962, 'Frozen Fish that Are "Seafresh"', *Food Process* **23**, 50.

Mitchell, R. S., Casmir, D. J., and Lynch, L. J.: 1961, 'The Maturometer – Instrumental Test and Redesign', *Food Technol.* **15**, 415.

Miyada, D. S. and Tappel, A. L.: 1956, 'Meat Tenderization. I: Two Mechanical Devices for Measuring Texture', *Food Technol.* **10**, 142.

Miyauchi, D. T.: 1963, 'Drip Formation in Fish. I: A Review of Factors Affecting Drip', *Fish Indust. Res.* **2**, 13.

Mohsenin, N. N.: 1963, 'A Testing Machine for Determining the Mechanical and Rheological Properties of Agricultural Products', *Penn. State Univ. Agr. Expt. Sta. Bull.* 701.

Mohsenin, N. N.: 1970a, *Physical Properties of Plant and Animal Materials*, Volume I – 'Structure, Physical Characteristics and Mechanical Properties', Gordon and Breach Science Publishers, New York, 734 pp.

Mohsenin, N. N.: 1970b, 'Application of Engineering Techniques to Evaluation of Texture of Solid Food Materials', *J. Texture Studies* **1**, 133.

Mohsenin, N. N. and Morrow, C. T.: 1968, 'Measurement of Viscoelastic Parameters in Food Materials', Soc. Chem. Ind. Monograph No. 27, London, p. 50.

Molnar, S.: 1968, 'Mechanical Simulation of Human Chewing Motions', *J. Dent. Res.* **47**, 559.

Morris, O. M.: 1925, 'Studies in Apple Storage', *Wash. Agr. Expt. Sta. Bull.* 193.

Morris, E. R.: 1965, 'Objective Tests for Use in the Technology of Compressed Foods', U.S. Army Natick Labs Tech. Rep. FD-26.

Morrow, C. T. and Mohsenin, N. N.: 1966, 'Consideration of Selected Agricultural Products as Viscoelastic Materials', *J. Food Sci.* **31**, 686.

Moskowitz, H. R.: 1972, 'Subjective Ideals and Sensory Optimization in Evaluating Perceptual Dimensions in Food', *J. Appl. Psychology* **56**, 60.

Moskowitz, H. R.: 1972, 'Scales of Subjective Viscosity and Fluidity of Gum Solutions', *J. Texture Studies* **3**, 89.

Moyer, J. C., Wilson, D. E., and Hand, D. B.: 1956, 'An Instrument for Evaluating the Texture of Fruits and Vegetables', *N.Y.S. Agr. Expt. Sta.*, Unpub. Ms., No. 1056 (see Drake, 1961).

Mühlemann, R. R.: 1960, 'Ten Years of Tooth-Mobility Measurements', *J. Periodont* **31**, 110.

Muller, H. G.: 1969, 'Mechanical Properties, Rheology, and Haptaesthesis of Foods', *J. Texture Studies* **1**, 38.

Murphy, Elizabeth F., True, Ruth H., and Hogan, J. M.: 1967, 'Detection Threshold of Sensory Panels for Mealiness of Baked Potatoes as Related to Specific Gravity Differences', *Am. Potato J.* **44**, 442.

Natrella, Mary G.: 1963, *Experimental Statistics*, National Bureau of Standards (U.S.), Handbook 91.

Nelmes, Brenda J. and Preston, R. D.: 1968, 'Wall Development in Apple Fruits: a Study of the Life History of a Parenchyma Cell', *J. Exp. Botany* **19**, 496.

Nemitz, G.: 1963, 'A New Device for Measuring the Course of Rigor Mortis in Fish and Mammalian Muscles' (in German), *Fleischwirtschaft* **15**, 18.

Oldfield, R. C.: 1960, 'Perception in the Mouth', in *Texture in Foods*, Monograph No. 7, Soc. Chem. Ind., London, p. 3.

Palmer, W. E.: 1962, 'Tenderness Testing Device', Canadian Patent 639 364.

Pangborn, R. M., Trabue, I. M., and Szczesniak, A. S.: 1972, 'Effect of Hydrocolloids on Oral Viscosity and Basic Taste Intensities', *J. Texture Studies* (in press).

Parkinson, C., Sherman, P., and Matsumoto, S.: 1970, 'Fat Crystals and the Flow Rheology of Butter and Margarine', *J. Texture Studies* **1**, 206.

Partmann, W.: 1971, 'Contractability of Muscle Fibres as a Possible Aid to Judging Changes in Frozen Meats', *J. Texture Studies* **2**, 328.

Patton, T. C.: 1969, 'Viscosity Profile of Typical Polysaccharides in the Ultra-Low Shear Rate Range', *Cereal Sci. Today* **14**, 178.

Paul, Pauline: 1949, 'Methods Used in the Study of Histological Structure of Meat', Proc. 2nd Ann. Reciprocal Meat Conf., p. 90.

Personius, Catherine J. and Sharpe, P. F.: 1938, 'Adhesion of Potato-Tuber Cells as Influenced by Temperature', *Food Res.* **3**, 513.

Peryam, D. R. and Pilgrim, F. J.: 1957, 'Hedonic Scale Method of Measuring Food Preferences', *Food Technol.* **11**, 9.

Pfaffmann, C.: 1939, 'Afferent Impulses from the Teeth from Pressure and Noxious Stimulation', *J. Physiol.* **97**, 207.

Pierson, A. and LeMagnen, J.: 1970, 'Study of Food Textures by Recording Chewing and Swallowing Movements', *J. Texture Studies* **1**, 327.

Pintauro, N. D. and Lang, R. E.: 1959, 'Graphic Measurement of Unmolded Gels', *Food Res.* **24**, 310.

Platt, W.: 1930, 'Staling of Bread', *Cereal Chem.* **7**, 1.

Platt, W. and Powers, R.: 1940, 'Compressibility of Bread Crumb', *Cereal Chem.* **17**, 601.

Pool, M. F.: 1967, 'Objective Measurements of Connective Tissue Tenacity of Poultry Meat', *J. Food Sci.* **32**, 550.

Pool, M. F. and Klose, A. A.: 1969, 'The Relation of Force to Sample Dimensions in Objective Measurements of Tenderness of Poultry Meat', *J. Food Sci.* **34**, 524.

Poulton, E. C.: 1968, 'The New Psychophysics: Six Models for Magnitude Estimation', *Psychol. Bull.* **69**, 1.

Prentice, J. H.: 1953, 'A Note on the Electrical Resistance and the Keeping Quality of Butter', *J. Dairy Res.* **20**, 327.

Prentice, J. H.: 1954, 'An Instrument for Estimating the Spreadability of Butter', *Lab. Pract.* **3**, 186.

Proctor, B. E., Davison, S., Malecki, G. J., and Welch, May: 1955, 'A Recording Strain Gage Denture Tenderometer for Foods, I: Instrument Evaluation and Initial Tests', *Food Technol.* **9**, 47.

Proctor, B. E., Davison, S., and Brody, A. L.: 1956a, 'A Recording Strain Gage Denture Tenderometer for Foods. II: Studies on the Masticatory Force and Motion, and the Force Penetration Relationship', *Food Technol.* **10**, 327.

Proctor, B. E., Davison, S., and Brody, A. L.: 1956b, 'A Recording Strain Gage Denture Tenderometer for Foods. III: Correlation with Subjective Tests and the Canco Tenderometer', *Food Technol.* **10**, 344.

Pyhälä, E.: 1935, 'Consistency Measurements According to Hj. Crusell', *Öle, Fette, Wachse* **19**, 13.

Quarmby, A. R.: 1971, CSIRO Food Research Report No. 44 (Available on request from CSIRO Tasmanian Regional Laboratory, Hobart, Tasmania, Australia).

Rasekh, J.: 1968, 'The Application of Headspace Gas-Liquid Chromatography for Measuring Quality of Fresh and Processed Vegetables', Ph.D. Thesis, Univ. Maryland.

Rasekh, J., Kramer, A., and Finch, R.: 1970, 'Objective Evaluation of Canned Tuna Sensory Quality', *J. Food Sci.* **35**, 417.

Reeve, R. M.: 1954a, 'Histological Survey of Conditions Influencing Texture in Potatoes. I: Effects of Heat Treatment on Structure', *Food Res.* **19**, 323.

Reeve, R. M.: 1954b, 'Histological Survey of Conditions Influencing Texture in Potatoes. II: Observations on Starch in Treated Cells', *Food Res.* **19**, 333.

Reeve, R. M.: 1970, 'Relationships of Histological Structure to Texture of Fresh and Processed Fruits and Vegetables', *J. Texture Studies* **1**, 247.

Reiner, M.: 1960, *Deformation, Strain, and Flow*, Lewis and Company, London, 347 pp.

Reiner, M. and Scott Blair, G. W.: 1967, 'Rheological Terminology', in *Rheology Theory and Applications*, Vol. IV (ed. by F. R. Eirich), Academic Press, New York.

Rich, A. D.: 1942, 'Methods Employed in Expressing the Consistency of Plasticized Shortenings', *Oil and Soap* **19**, 54.

Richards, J. F. and Staley, L. M.: 1967, 'The Relationships Between Crushing Strength, Deformation and Other Physical Measurements of the Hen's Egg', *Poultry Sci.* **46**, 430.

Robson, A. H.: 1966, 'The Measurement of Cake Crumb Strength', *J. Food. Technol.* **1**, 291.

Romani, R. J., Jacob, F. C., and Sprock, C. M.: 1962, 'Studies on the Use of Light Transmission to Assess the Maturity of Peaches, Nectarines and Plums', *Proc. Am. Soc. Hort. Sci.* **80**, 220.

Rowlands, D. G.: 1964, 'Measurement of Varietal Differences in Firmness of Brussel Sprouts', *Hort. Res.* **3**, 102.

Sato, Y. and Nakayama, T.: 1970, 'Discussion of the Binding Quality of Minced Meats Based on Their Rheological Properties Before and After Heating, *J. Texture Studies* **1**, 309.

Sato, Y. and Teruo, N.: 1970, 'Discussion of the Binding Quality of Minced Meats Based on Their Rheological Properties Before and After Heating', *J. Texture Studies* **1**, 309.

Satorius, Mary J. and Child, Alice M.: 1938, 'Effect of Coagulation on Press Fluid, Shear Force, Muscle Cell Diameter and Composition of Beef Muscle', *Food Res.* **3**, 619.

Schachat, R. E. and Nacci, A.: 1960, 'Transistorized Bloom Gelometer', *Food Technol.* **14**, 117.

Scherr, H. J. and Witnauer, L. P.: 1967, 'The Application of a Capillary Extrusion Rheometer to the Determination of the Flow Characteristics of Lard', *J. Am. Oil Chem. Soc.* **44**, 275.

Schlichting, Hermann: 1968, *Boundary Layer Theory* (translated by J. Kestin), McGraw-Hill Book Co., New York, 6th ed.

Schmidt, A. X. and Marlies, C. A.: 1948, *Principles of High-Polymer Theory and Practices*, McGraw-Hill Book Co., New York.

Schomer, H. A. and Olsen, K. L.: 1962, 'A Mechanical Thumb for Determining Firmness of Apples', *Proc. Am. Soc. Hort. Sci.* **81**, 61.

Schutz, H. G.: 1965, 'A Food Action Rating Scale for Measuring Food Acceptance', *J. Food Sci.* **30**, 365.

Scott Blair, G. W.: 1953, *Foodstuffs – Their Elasticity, Fluidity, and Consistency*, North-Holland Pub. Co., Amsterdam.

Scott Blair, G. W.: 1958, 'Rheology in Food Research', *Adv. Food Res.* **8**, 1.

Scott Blair, G. W.: 1966, 'The Subjective Assessment of the Consistency of Materials in Relation to Physical Measurement', *J. Soc. Cosmetic Chem.* **17**, 45.

Scott Blair, G. W.: 1968, Comments made during symposium on sensory evaluation at Swedish Institute for Food Preservation Research, Goteborg, Sweden, September.

Scott Blair, G. W.: 1969, 'Rheology: a Brief Historical Survey', *J. Texture Studies* **1**, 14.

Scott Blair, G. W. and Burnett, J.: 1957, 'An Apparatus for Measuring the Elastic Properties of Very Soft Gels', *Laboratory Practice* **6**, 570.

Scott Blair, G. W. and Reiner, M.: 1957, *Agricultural Rheology*, Routledge and Kegan Paul, Ltd., London, 222 pp.

Segars, R. and Kapsalis, J. G.: 1971, Unpublished data.

Shama, F. and Sherman, P.: 1966, 'The Texture of Ice Cream. II: Rheological Properties of Frozen Ice Cream', *J. Food Sci.* **31**, 699.

Shama, F. and Sherman, P.: 1968, 'An Automated Parallel Plate Viscoelastometer for Studying the Rheological Properties of Solid Food Materials', in *Rheology and Texture of Foodstuffs*, Soc. Chem. Ind. Monograph No. 27, London, p. 77.

Shama, F. and Sherman, P.: 1970, Unpublished data.

Shama, F. and Sherman, P.: 1973a, 'Identification of Stimuli Controlling the Sensory Evaluation of Viscosity. II: Oral Methods', *J. Texture Studies* **4**, 111.

Shama, F. and Sherman, P.: 1973b, 'Variation in Stimuli Associated with Oral Evaluation of the Viscosities of Glucose Solutions', *J. Texture Studies* **4**, 254.

Shama, F., Parkinson, C., and Sherman, P.: 1973, 'Identification of Stimuli Controlling the Sensory Evaluation of Viscosity, I: Non Oral Methods', *J. Texture Studies* **4**, 102.

Sharma, M. G. and Mohsenin, N. N.: 1970, 'Bio-Mechanical Investigation of a Fruit Subjected to Hydrostatic Pressure', *Proc. Fifth Int. Cong. Rheology* **2**, 667.

Sherman, P.: 1965, 'The Texture of Ice Cream, Part 1', *J. Food Sci.* **30**, 201.

Sherman, P.: 1966, 'The Texture of Ice Cream, III: Rheological Properties of Mix and Melted Ice Cream', *J. Food Sci.* **31**, 707.

Sherman, P.: 1969, 'A Texture Profile of Foodstuffs Based on Well-Defined Rheological Properties', *J. Food Sci.* **34**, 458.

Sherman, P.: 1970, *Industrial Rheology with Particular Reference to Foods, Pharmaceuticals and Cosmetics*, Academic Press, New York.

Sherman, P.: 1971, Personal correspondence.

Sherman, P.: 1972, 'Structure and Textural Properties of Foods', *Food Technol.* **26**, 69.

Shewfelt, A. L.: 1965, 'Changes and Variations in the Pectic Constitution of Ripening Peaches as Related to Product Firmness', *J. Food Sci.* **30**, 573.

Showalter, R. K.: 1961, 'Specific Gravity, Weight, and Solids Relationships in Watermelons', *Proc. Florida State Hort. Soc.* **74**, 268.

Simon, S., Field, J. C., Kramlich, W. E., and Tauber, F. W.: 1965, 'Factors Affecting Frankfurter Texture and a Method of Measurement', *Food Technol.* **19**, 410.

Sjöberg, L.: 1966, 'A Method for Sensation Scaling Based on an Analogy Between Perception and Judgment', *Percept. Psychophysics* **1**, 131.

Sjöberg, L.: 1971, 'Three Models for the Analysis of Subjective Ratios', *Scand. J. Psychol.* **3**, 217.

Slater, L. E.: 1954, 'Rheology Opens Food Frontiers', *Food. Eng.* **26**, 74.

Smith, H. R.: 1947, 'Objective Measurements of Quality in Foods', *Food Technol.* **1**, 345.

Smith, E. E.: 1957, 'Tenderness and Chemical Changes in Muscle Fibers During Cooking', M.S. Thesis, Purdue University, La Fayette, Ind.

Smith, Jean W. and Kramer, A.: 1972, 'Palatability and Nutritive Value of Fresh, Canned, and Frozen Collard Greens', *J. Amer. Soc. Hort. Sci.* **97**, 161.

Smith, H. and Rose, A.: 1963, 'Subjective Responses in Process Investigation', *Ind. Eng. Chem.* **55**, 25.

Smith, J. R., Smith, T. L., and Tschoegl, N. W.: 1970, 'Rheological Properties of Wheat Flour Doughs. III: Dynamic Shear Modulus and Its Dependence on Amplitude, Frequency, and Dough Composition', *Rheologica Acta* **9**, 239.

Snedecor, G. W. and Cochran, W. G.: 1967, *Statistical Methods*, Iowa State University Press.

Sokal, R. R. and Rohlf, F. J.: 1969, *Biometry; the Principles and Practice of Statistics in Biological Research*, W. H. Freeman and Co., San Francisco.

Sokolnikoff, I. S.: 1956, *Mathematical Theory of Elasticity*, McGraw-Hill Book Co., New York, 2nd ed., 476 pp.

Somers, G. F.: 1965, 'Viscoelastic Properties of Storage Tissues from Potato, Apple and Pear', *J. Food Sci.* **30**, 922.

Somers, G. F.: 1966, 'The Bending of Potato-Tuber Slices Mounted as Cantilever Beams', *J. Exper. Botany* **17**, 27.

Sone, T.: 1961, 'The Rheological Behavior and Thixotropy of a Fatty Plastic Body', *J. Phys. Soc. Japan* **16**, 961.

Southorn, W. A.: 1960, 'The Torsiometer – an Instrument for the Study of Gels Considered as Elastic Solids', *J. Sci. Instr.* **37**, 292.

Spencer, J. V., Jacobson, M., and Kimbress, J. T.: 1962, 'Recording Strain Gage Shear Apparatus', *Food Technol.* **16**, 113.

Sperring, Doris, Platt, D. W. T., and Hiner, R. L.: 1959, 'Tenderness in Beef Muscle as Measured by Pressure', *Food Technol.* **12**, 155.

Stanley, L. M.: 1970, 'Materials for Standardizing the FMC Tenderometer', *J. Can. Inst. Food Technol.* **3**, 116.

Steel, R. G. D. and Torrie, K. T.: 1960, *Statistical Analysis*, J. Wiley & Sons, Inc., New York.

Sterling, C.: 1955, 'Effect of Moisture and High Temperature on Cell Walls in Plant Tissues', *Food Res.* **20**, 474.

Sterling, C.: 1963, 'Texture and Cell Wall Polysaccharides in Foods', *Recent Advances in Food Science* (ed. by J. M. Leitch and D. N. Rhodes), Butterworths, London, **3**, 259.

Stevens, S. S.: 1951, 'Mathematics, Measurement, and Psychophysics', in *Handbook of Experimental Psychology* (ed. by S. S. Stevens, Wiley, New York, p. 25.

Stevens, S. S.: 1953, 'On the Brightness of Lights and the Loudness of Sounds', *Science* **118**, 576.

Stevens, S. S.: 1960, 'The Psychophysics of Sensory Function', *American Scientist* **48**, 226.

Stevens, S. S.: 1966a, 'Matching Functions Between Loudness and Ten Other Continua', *Perception Psychophysics* **1**, 5.

Stevens, S. S.: 1966b, 'On the Operation Known as Judgement', *American Scientist* **54**, 385.

Stevens, S. S. and Greenbaum, Hilda B.: 1966, 'Regression Effect in Psychophysical Judgment' *Perception Psychophysics* **1**, 439.

Stevens, S. S. and Guirao, Miguelina: 1963, 'Subjective Scaling of Length and Area and the Matching of Length to Loudness and Brightness', *J. Exper. Psychology* **66**, 177.

Stevens, S. S. and Guirao, Miguelina: 1964, 'Scaling of Apparent Viscosity', *Science* **144**, 1157.

Stevens, S. S. and Harris, Judith R.: 1962, 'The Scaling of Subjective Roughness and Smoothness' *J. Exper. Psychology* **64**, 489.

Stier, Elizabeth F.: 1970, 'Chew a Chip & Tell', *Food Technol.* **24**, 46.

Stinson, C. G. and Huck, M. B.: 1969, 'A Comparison of Four Methods for Pastry Tenderness Evaluation', *J. Food Sci.* **34**, 537.

Suppes, P.: 1951, 'A Set of Independent Axioms for Extensive Quantities', *Portugaliae Mathematica* **10**, 163.

Suppes, P. and Zinnes, J. L.: 1963, 'Basic Measurement Theory', *Handbook of Mathematical Psychology*, (ed. by R. D. Luce, R. R. Bush and E. Galanter), Vol. I. John Wiley and Son, New York, p. 1.

Swartz, Verona W.: 1938, 'Two Further Simple Objective Tests for Judging Cake Quality', *Cereal Chem.* **15**, 247.

Szczesniak, Alina S.: 1963a, 'Objective Measurement of Food Texture', *J. Food Sci.* **28**, 410.

Szczesniak, Alina S.: 1963b, 'Classification of Textural Characteristics', *J. Food Sci.* **28**, 385.

Szczesniak, Alina S.: 1966, 'Texture Measurements', *Food Technol.* **20**, 1292.

Szczesniak, Alina S.: 1968, 'Correlations Between Objective and Sensory Texture Measurements', *Food Technol.* **22**, 981.

Szczesniak, Alina S.: 1971, 'Consumer Awareness of Texture and of Other Food Attributes, II', *J. Texture Studies* **2**, 196.

Szczesniak, Alina S.: 1973a, 'Instrumental Methods of Texture Measurement', this volume, p. 71.

Szczesniak, Alina S.: 1973b, 'Indirect Methods of Objective Texture Measurements', this volume, p. 109.

Szczesniak, Alina S. and Bourne, M. C.: 1969, 'Sensory Evaluation of Food Firmness', *J. Texture Studies* **1**, 52.

Szczesniak, Alina S. and Farkas, Elizabeth: 1962, 'Objective Characterization of the Mouthfeel of Gum Solutions', *J. Food Sci.* **27**, 381.

Szczesniak, Alina S. and Kleyn, D. H.: 1963, 'Consumer Awareness of Texture and Other Food Attributes', *Food Technol.* **17**, 74.

Szczesniak, Alina S. and Smith, Bertha J.: 1969, 'Observations on Strawberry Texture. A Three-Pronged Approach', *J. Texture Studies* **1**, 65.

Szczesniak, Alina S. and Torgeson, Kathryn W.: 1965, 'Methods of Meat Texture Measurement Viewed from the Background of Factors Affecting Tenderness', *Adv. Food Res.* **14**, 33.

Szczesniak, Alina S., Brandt, Margaret A., and Friedman, H. H.: 1963a, 'Development of Standard

Rating Scales for Mechanical Parameters of Texture and Correlation Between the Objective and Sensory Methods of Texture Evaluation', *J. Food Sci.* **28**, 397.

Szczesniak, Alina S., Sloman, Katherine, Brandt, Margaret, and Skinner, Elaine Z.: 1963b, 'Objective Measurement of Texture of Fresh and Freeze-Dehydrated Meats', *Proc. 15th Res. Conf. Am. Meat Inst. Foundation* (circular No. 74), 121.

Szczesniak, Alina S., Humbaugh, P. R., and Block, H. W.: 1970, 'Behavior of Different Foods in the Standard Shear Compression Cell of the Shear Press and the Effect of Sample Weight on Peak Area and Maximum Force', *J. Texture Studies* **1**, 356.

Tanaka, M., de Man, J. M., and Voisey, P. W.: 1971, 'Measurement of Textural Properties of Foods with a Constant Speed Cone Penetrometer', *J. Texture Studies* **2**, 306.

Tarr, L. W.: 1926, 'Fruit Jellies. III: Jelly Strength Measurements', *Delaware Agr. Expt. Sta. Bull.* 142.

Tarver, Mae-Goodwin and Schenck, Anna May: 1958, 'Statistical Development of Objective Quality Scores for Evaluating the Quality of Food Products. Development of Scoring Scales', *Food Technol.* **12**, 127.

Timoshenko, S. P. and Goodier, J. N.: 1970, *Theory of Elasticity*, McGraw-Hill, New York, 3rd ed.

Toda, J., Wada, T., Yasumatsu, K., and Ishii, K.: 1971, 'Application of Principal Component Analysis to Food Texture Measurements', *J. Texture Studies* **2**, 207.

Torgerson, W. S.: 1965, 'Multidimensional Scaling of Similarity', *Psykometrika* **30**, 379.

Townsend, C. T., Somers, I. R., Lamb, F. C., and Olson, N. A.: 1956, 'A Laboratory Manual for the Canning Industry', National Canners Assoc. Res. Lab., Washington, D.C., p. 19.

Truelson, T. A.: 1963, 'Radiation Pasteurization of Fresh Fruits and Vegetables', *Food Technol.* **17**, 336.

Twigg, B. A.: 1960, 'Food Textural Measuring Instruments and Their Standardization and Calibration', Paper presented at 20th Annual Meeting, Institute of Food Technologists, San Francisco, May 15–19.

Twigg, B. A., Kramer, A., Falen, H. N., and Southerland, F. L.: 1956, 'Objective Evaluation of the Maturity Factor in Processed Sweet Corn', *Food Technol.* **10**, 171.

U.S. Department of Agriculture, Consumer and Marketing Service, *Grades and Standards of Quality*, available from the different commodity branches, Washington, D.C.

Van Wazer, J. R., Lyons, J. W., Kin, K. Y., and Colwell, R. E.: 1963, *Viscosity and Flow Measurement: A Laboratory Handbook of Rheology*, Interscience Pub., New York.

Virgin, H. I.: 1955, 'A New Method for the Determination of the Turgor of Plant Tissues', *Physiologia Plantarum* **8**, 954.

Voisey, P. W.: 1971, 'Modernization of Texture Instrumentation', *J. Texture Studies* **2**, 129.

Voisey, P. W. and Emmons, D. B.: 1966, 'Modifications of the Curd Firmness Test for Cottage Cheese', *J. Dairy Sci.* **49**, 93.

Voisey, P. W. and Hansen, H.: 1967, 'A Shear Apparatus for Meat Tenderness Evaluation', *Food Technol.* **21**, 355.

Voisey, P. W. and Hunt, J. R.: 1967, 'Relationship Between Applied Force, Deformation of Egg Shells and Fracture Force', *J. Agr. Eng. Res.* **12**, 1.

Voisey, P. W. and Nonnecke, I. L.: 1971, 'Measurement of Pea Tenderness. I: An Appraisal of the F.M.C. Pea Tenderometer', *J. Texture Studies* **2**, 348.

Voisey, P. W. and Nunes, A.: 1968, 'An Electronic Recording Amylograph', *J. Can. Inst. Food Technol.* **1**, 128.

Volodkevich, N. N.: 1938, 'Apparatus for Measurement of Chewing Resistance or Tenderness of Foodstuffs', *Food Res.* **3**, 221.

von Elbe, J. H. and Johnson, C. E.: 1971, 'Sales Pattern Changing – Color May Be a Factor', *Canner/Packer* **140**, 10.

Warner, K. F.: 1927, 'A Study of the Factors Which Influence the Quality and Palatability of Meat', U.S. Dept. Agr. Nat. Coop. Proj., Coop. Bur. Animal Ind. Rev. ed.

Wegener, J. B., Baer, Beverly H., and Rogers, P. D.: 1951, 'Improving Quality of Frozen Strawberries with Added Colloids', *Food Technol.* **5**, 76.

Wen, P. R. and Mohsenin, N. N.: 1970, 'Measurement of Dynamic Viscoelastic Properties of Corn Horny Endosperm', *J. Materials* **5**, 856.

White, R. K. and Mohsenin, N. N.: 1967, 'Apparatus for Determination of Bulk Modulus and Compressibility of Materials', *Trans. Am. Soc. Agr. Eng.* **10**, 670.

Whitehead, R. C.: 1970, 'Fat Analysis of Boned Meat by the Specific Gravity Method', *Food Technol.* **24**, 469.

Whitehead, J. and Sherman, P.: 1967, 'Texture of Ice Cream. IV: The Influence of Fat Content and Coagulated Fat on the Structure of Melted Ice Cream', *Food Technol.* **21**, 1521.

Whittenberger, R. T.: 1951, 'Measuring the Firmness of Cooked Apple Tissues', *Food Technol.* **5**, 17.

Whittenberger, R. T. and Nutting, G. C.: 1957, 'Effect of Tomato Cell Structures on Consistency of Tomato Juice', *Food Technol.* **11**, 19.

Whittenberger, R. T. and Nutting, G. C.: 1958, 'High Viscosity of Cell Wall Suspensions Prepared from Tomato Juice', *Food Technol.* **12**, 420.

Wine, R. L.: 1964, *Statistics for Scientists and Engineers*, Prentice-Hall, Inc., Englewood Cliffs, New Jersey.

Winkler, C. A.: 1939, 'Tenderness of Meat. I: A recording Apparatus for Its Estimation, and Relation Between pH and Tenderness', *Can. J. Res.* **17** (D), 8.

Wolodkewitsch, N.: 1956, 'Zur Methodik der Festigkeits Messungen an Lebensmitteln', *Z. Lebensm. Untersuch. Forsch.* **103**, 261.

Wood, F. W.: 1968, 'Psychophysical Studies on the Consistency of Liquid Foods', in *Rheology and Texture of Foodstuffs*, Soc. Chem. Ind. Monograph, No. 27, London, p. 40.

Woodmansee, C. W., McClendon, J. H., and Somers, G. F.: 1959, 'Chemical Changes Associated with the Ripening of Apples and Tomatoes', *Food Res.* **24**, 503.

Worner, H. K. and Anderson, M. N.: 1944, 'Biting Force Measurements on Children', *Austral. J. Dent.* **48**, 1.

Yoshida, M.: 1968, 'Dimensions of Tactual Impressions. 1 and 2', *Jap. Psychol. Res.* **10**, 123, 157.

Yoshikawa, S., Nishumaru, S., and Yoshida, M.: 1970, 'Collection and Classification of Words. for Description in Food Texture. I: Collection of Words. II: Texture Profiles. III: Classification by Multivariate Analysis', *J. Texture Studies* **1**, 437, 443, 452.

Young, G. and Householder, A. S.: 1938, 'Discussion of a Set of Points in Terms of Their Mutual Distances', *Psykometrika* **3**, 19.

Zoerb, G. C. and Hall, C. W.: 1960, 'The Mechanical and Rheological Properties of Grains', *J. Agr. Eng. Res.* **5**, 83.